THERMAL SOLID WASTE UTILISATION IN REGULAR AND INDUSTRIAL FACILITIES

ENVIRONMENTAL SCIENCE RESEARCH

Series Editior:

Herbert S. Rosenkranz
Department of Environmental and Occupational Health
Graduate School of Public Health
University of Pittsburgh
130 DeSoto Street
Pittsburgh, Pennsylvania

Founding Editor:

Alexander Hollaender

Recent Volumes in this Series

Volume 47—GLOBAL ENERGY STRATEGIES: Living with Restricted Greenhouse Gas Emissions
Edited by James C. White

Volume 48—GLOBAL ATMOSPHERIC-BIOSPHERIC CHEMISTRY
Ronald G. Prinn

Volume 49—BIODEGRADATION OF NITROAROMATIC COMPOUNDS
Edited by Jim C. Spain

Volume 50—BIOMONITORS AND BIOMARKERS AS INDICATORS OF ENVIRONMENTAL CHANGE: A Handbook
Edited by Frank M. Butterworth, Lynda D. Corkum, and Judith Guzmán-Rincón

Volume 51—CHEMISTRY FOR THE PROTECTION OF THE ENVIRONMENT 2
Edited by Lucjan Pawłowski, William J. Lacy, Christopher G. Uchrin, and Marzenna R. Dudzińska

Volume 52—PRODUCED WATER 2: Environmental Issues and Mitigation Technologies
Edited by Mark Reed and Ståle Johnsen

Volume 53—EVALUATING CLIMATE CHANGE ACTION PLANS: National Actions for International Commitment
Edited by James C. White

Volume 54—BIOTECHNOLOGY IN THE SUSTAINABLE ENVIRONMENT
Edited by Gary S. Sayler, John Sanseverino, and Kimberly L. Davis

Volume 55—CHEMISTRY FOR THE PROTECTION OF THE ENVIRONMENT 3
Edited by Lucjan Pawłowski, Marjorie A. Gonzales, Marzenna R. Dudzińska, and William J. Lacy

Volume 56—BIOMONITORS AND BIOMARKERS AS INDICATORS OF ENVIRONMENTAL CHANGE 2
Edited by Frank M. Butterworth, Amara Gunatilaka, and Maria Eugenia Gonsebatt

Volume 57—PERCHLORATE IN THE ENVIRONMENT
Edited by Edward Todd Urbansky

Volume 58—THERMAL SOLID WASTE UTILISATION IN REGULAR AND INDUSTRIAL FACILITIES
Edited by Lucjan Pawłowski, Marzenna R. Dudzińska, and Marjorie A. Gonzalez

THERMAL SOLID WASTE UTILISATION IN REGULAR AND INDUSTRIAL FACILITIES

Edited by

Lucjan Pawłowski

Marzenna R. Dudzińska
Lublin University of Technology
Lublin, Poland

and

Marjorie A. Gonzalez
Lawrence Livermore National Laboratory
Livermore, California

Springer Science+Business Media, LLC

Proceedings of the International Workshop on Thermal Solid Waste Utilisation in Regular and Industrial Facilities, held November 28–30, 1999, in Kazimierz Dolny, Poland

ISBN 978-1-4613-6891-5   ISBN 978-1-4615-4213-1 (eBook)
DOI 10.1007/978-1-4615-4213-1

Originally published by Kluwer Academic / Plenum Publishers in 2000
Softcover reprint of the hardcover 1st edition 2000
http://www.wkap.nl

10 9 8 7 6 5 4 3 2 1

A C.I.P. record for this book is available from the Library of Congress

PREFACE

Human civilisation produces large amounts of wastes. In the past, most of the waste consisted of biodegradable organic matter that was managed by disposal methods.

In more recent years, industrial and urban growth and development have driven not only positive changes, such as improvements in the standard of living, but have also resulted in the generation of growing volumes of non-degradable wastes. At the same time, landfill siting has become more and more difficult, principally caused by a shortage of available land near urban areas, the potential for the wastes to contaminate ground water, and community resistance to waste disposal sites being in their backyards. As a result, technologies related to recycling and material reuse are being favoured over land disposal. While the material reuse of flammable wastes may not be the most environmentally friendly option, these solutions might be preferable for wastes of little or no calorific value.

Paper recycling generates large amounts of secondary contaminants, and the recycled final product is generally a lower grade paper product since the cellulose fibbers are shorted in the process. However, at the same time, a lot of efforts is being put into biomass fuel production (cultivation) projects, based on specially established plantations.

Assuming that waste paper incineration allows energy recovery, one can treat waste paper as biomass fuel. Additionally, this biomass, derived from the quickly cultivated trees, is used as a paper product prior to its use as a fuel.

A similar situation exists for plastic incineration. In most cases, plastic recycling causes more secondary pollution than that created during the primary production from crude oil. Also the quality of the second-generation product is lower. So, recognition of plastics as raw materials for energy production might be more friendly for the environment. Energy recovery from plastics during incineration also reduces consumption of fossil fuels. Oil or coal can be used for polymer production with lower atmospheric emission.

Realisation of such a concept requires economic incentives and the pairing of two complimentary processes: waste incineration and energy recovery. From this perspective, co-incineration processes, including cement production, are the most economically profitable. Clinker kiln allows the incineration of a broad spectrum of wastes, with energy recovery and neutralisation of acidic and other emissions managed in the process. Additionally, generated ashes are built into the clinker structure. This process is considered to be environmentally friendly, as it does not cause secondary emissions.

The editors acknowledge the Polish Ministry for National Education and the State Committee for Scientific Research for sponsoring publication of this volume.

L. Pawłowski, M. R. Dudzińska, and M. A. Gonzalez

CONTENTS

NEUTRALISATION OF WASTES IN A CEMENT KILN

Lucjan Pawłowski
Institute of Environmental Protection Engineering, Lublin University of Technology, 40 Nadbystrzycka Street, 20-618 Lublin, Poland

Key words: waste, incineration, cement kiln, clinker production

Abstract: Application of a cement kiln for incineration of wastes is reviewed and impact of incineration on the environment is presented.

1. INTRODUCTION

Humankind has exerted a dominant influence over the environment since the dawn of history, and continues to do so today. There has been, is and will be no civilisation without the generation of greater or lesser amounts of waste, it is just that some of this may be reused in one form or another.

Nevertheless, even with partial recycling, the dumping of the remainder requires a large and ever greater amount of space – space that must be excluded from use for many years since the natural processes of degradation are very slow. It is worth noting, that even the organic matter that is a cob of corn did not undergo decomposition at a landfill site over several tens of years.

Even assuming the total recycling of metal, glass, paper and compostable organic matter, there will remain enormous amounts of municipal refuse, to say nothing of the industrial wastes requiring neutralisation or dumping.

The dumping of wastes is a stop-gap solution, since it does not fully eliminate secondary emissions to the environment, e.g. of the dioxins so widely-associated with combustion, but in fact also capable of being released by biochemical processes of waste decomposition.

Against this background, incineration may be considered a relatively safe technique, providing that it is done in closely monitored conditions. The course of the incineration process in even conventional incinerators is associated with a marked reduction in the volume of wastes, while the ashes produced may be dumped safely.

Incineration in connection with the recovery of energy may further allow for the reduced consumption of fossil fuels. There is thus no justification for a situation in which each recycling of wastes, even if actually associated with serious secondary emission of pollutants, is said to represent a desirable recycling of those wastes, while the recovery of energy from incineration continues to provoke controversy.

In the light of all this, the combustion of wastes in cement kilns may be looked upon as one of the very attractive measures available. In this case, the energy obtained can be used in the processing of clinker, thereby reducing the consumption of hard coal. In addition, it is easy to neutralise the products of incineration, which are the ashes permanently sintered in the structure of the clinker, as well as acidic components (SO_2 and NO_x) of the combustion gases, which react with the alkaline particulates and are retained in electrofilters.

2. A CHARACTERISATION OF THE PERFORMANCE OF A CEMENT KILN

The process of burning clinker has been presented schematically in Figure 1. Fuel is introduced to the rotary kiln from the lower, so-called "hot" end, most often in the form of coal dust. In turn, introduced at the most elevated end are raw materials in the form of a mixture of marl and limestone, as well as a certain amount of material containing iron compounds.

This mixture undergoes transformations as it passes through successive temperature zones. The most important of these are:

- the calcination ongoing at temperatures in the range 800-1200°C and entailing the breakdown of limestone with the production of large amounts of CO_2;
- the sintering ongoing at temperatures in the range 1400-1500°C and entailing the reaction of the oxides Al_2O_3, CaO, SiO_2 and Fe_2O_3 with the formation of what is known as clinker.

These are the two important processes from the point of view of the application of a cement kiln for neutralisation.

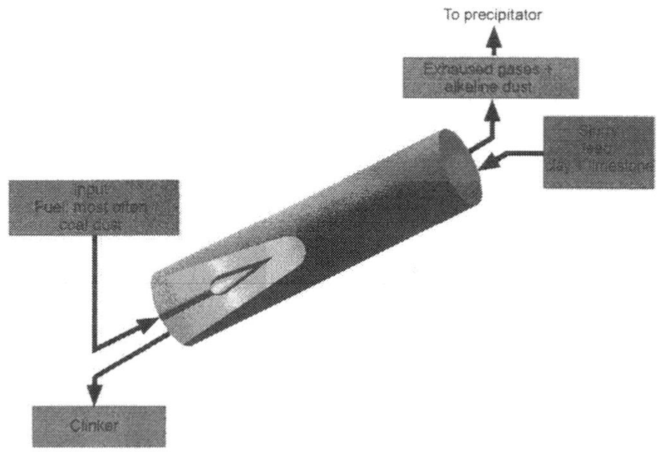

Figure 1. Flowsheet depicting the manufacturing process of clinker.

3. CHARACTERISTIC OF NEW PROCESSES

The processes may be divided into three groups:

- the burning of wastes of a certain calorific value linked with the production of heat;
- the sintering of mineral wastes combined with use of such components as CaO, Al_2O_3, Fe_2O_3 and SiO_2 as substitute for raw materials and linked with the permanent immobilisation of the metal oxides they contain;
- the immobilisation of metal oxides during the sintering process.

Each group of wastes should be introduced to a different section of the cement kiln (Fig. 2).

Clinker production is a high-temperature process. The temperature of the burning material in the sintering zone is in the range 1400-1500°C, while that of the combustion gases exceeds 1900°C. The large mass (1000-5000 Mg) of material passing through the kiln in an hour ensures a high degree of thermal stability, denoting ideal internal conditions for the complete breakdown of even the most chemically-resistant compounds like PCBs, providing that the wastes are introduced to the kiln in the appropriate way.

3.1 The Introduction of Wastes into a Cement Kiln

Several ways of introducing wastes into the kiln have been elaborated.

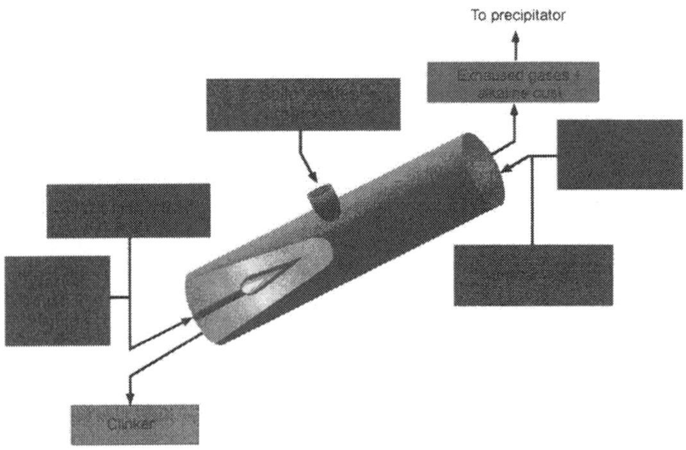

Figure 2. Characteristic of waste introduction into a cement kiln.

3.2 Liquid Wastes

Liquid wastes may be introduced into the kiln with the aid of a lance (Fig. 3).

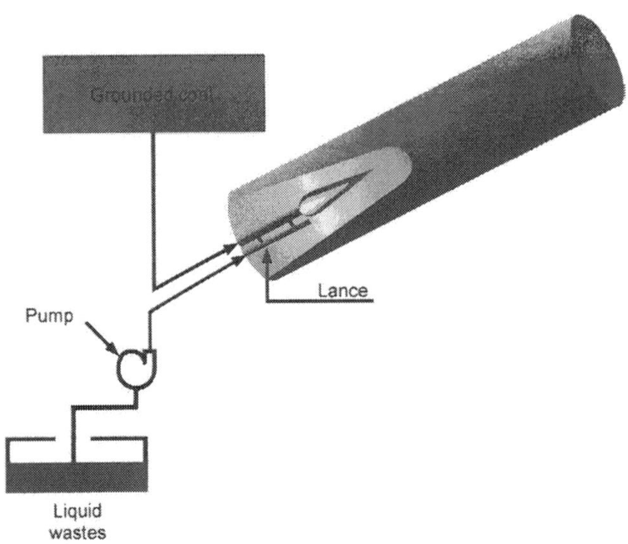

Figure 3. Characteristic of introducing liquid wastes into the kiln.

The lance is suspended beneath the coal-dust burner. The liquid waste is injected at a point where the temperature is of about 1500°C, at which there is total mineralisation of even such durable compounds as PCBs.

3.3 Solid Wastes

Solid wastes may be in dusty, ground or unground form.

Dusty wastes are mixed with coal dust and fed directly to the flame zone, i.e. that at a temperature of more than 1900°C. Municipal refuse may be included in this group, and we have elaborated methods for its grounding and supply with coal dust. We already have experimental equipment available that can process 400 kg of municipal waste per hour.

The second system for supplying doses of waste is the sluice in the kiln mantle. The principle by which this works is illustrated in Fig 4. With the aid of the experimental setup built in accordance with our patent it is possible to introduce about 1.2 Mg of wastes per hour into the section of the kiln at a temperature of 1250°C.

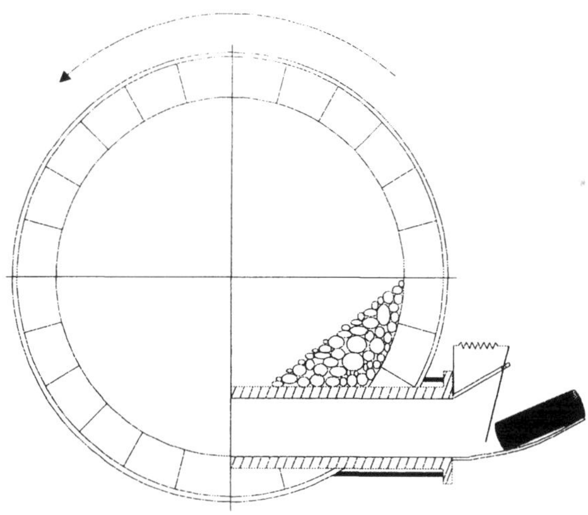

Figure 4. Illustration of sluice.

Another installation by which wastes my be introduced into a kilns is the so-called pneumatic discharge (Figs. 5 and 6), which allows for the shooting-in of three 10-15 kg charges per minute, i.e. 1.8 to 2.7 Mg per hour. The charges are introduced into the section of the kiln at 1400°C.

Figure 5. General view of the pneumatic discharge.

Figure 6. System introducing compressed air into the pneumatic discharge.

4. A CHARACTERISATION OF THE COMBUSTION PROCESS

Reference can be made to three different types of waste:

4.1 Wastes Comprising Hydrocarbons of General Formula $C_xH_yO_z$

In the case of these wastes, the combustion process can be depicted by the following equation of the reaction:

$$C_xH_yO_z + O_2 \rightarrow CO_2 + H_2O$$

The above products are the only ones and it is worth noting that - per unit of heat generated - the amount of carbon dioxide released in the burning of wastes of this type is smaller than that generated as an equivalent amount of coal is burnt. This implies that the combustion of organic wastes in a cement kiln actually leads to a decrease of carbon dioxide emissions.

4.2 Hydrocarbon Wastes also Containing Sulphurous Compounds of the General Formula $C_xH_yO_zS_n$

Here, the combustion process is as characterised by the following general equation of the reaction:

$$C_xH_yO_zS_n \rightarrow CO_2 + H_2O + SO_2$$

In this case, sulphur dioxide is generated, as with the burning of coal with a higher sulphur content. However, the alkaline dusts of a electric precipitator reacts with SO_2 and therefore, the sulphur dioxide does not reach the environment, but reacts with the CaO of the dusts in according to equation of the reaction:

$$CaO + SO_2 \rightarrow CaSO_3$$

The calcium sulphite generated is retained in the electrofilter.

4.3 Chlorinated Hydrocarbon (PVC, PCB) Having General Formula $C_xH_yO_zCl_n$

The process by which these wastes are combusted is illustrated by the general equation of the reaction:

$$C_xH_yO_zCl_n + O_2 \rightarrow CO_2 + H_2O + HCl$$

In this case, as with the combustion of compounds containing sulphur, the gaseous product (this time HCl) is capable to react with CaO in according to the equation of the reaction:

$$CaO + 2HCl \rightarrow CaCl_2 + H_2O$$

The calcium chloride generated is retained in the electrofilter. Products of the combustion of pesticides in laboratory conditions are present in Table 1.

Table 1. A characterisation of the products obtained from the combustion of pesticides (amounts calculated from the balance of chemical reactions).

Pesticide	Products of combustion (kg/kg)					
	CO_2	H_2O	SO_2	NO	$CaCl_2$	$Ca_3(PO_4)_2$
$C_{12}H_8OCl_6$	1.40	0.19	-	-	0.87	-
$C_9H_6O_3SCl_6$	0.98	0.13	0.16	-	0.82	-
$C_9H_{10}Cl_8$	0.99	0.23	-	-	1.10	-
$C_{14}H_9Cl_5$	1.75	0.23	-	-	0.78	-
$C_6H_6Cl_6$	0.92	0.19	-	-	0.38	-
$C_{16}H_{15}O_2Cl_3$	2.05	0.39	-	-	0.48	-
C_6Cl_6	0.94	-	-	-	1.17	-
$C_6H_5Cl_2N$	0.90	-	-	0.10	0.94	-
$C_6H_{14}O_3NS_2P$	1.09	0.52	0.53	0.12	-	0.64
$C_8H_{16}O_4S_2P$	1.30	0.53	0.47	-	-	0.57
$C_6H_5O_2S_3P$	1.07	0.55	0.78	-	-	0.63
$C_{10}H_{15}O_3SP$	1.79	0.55	0.26	-	-	0.63
$C_6H_{15}O_2S_3P$	1.07	0.55	0.55	-	-	0.63
$C_5H_{13}O_3S_3P$	0.89	0.47	0.77	-	-	0.63
$C_9H_{12}O_5NSP$	1.43	0.39	0.23	0.11	-	0.56
$C_4H_8O_2Cl_3P$	0.79	0.32	-	-	0.74	0.69
$C_{10}H_{19}O_6S_2P$	1.33	0.52	0.39	-	-	0.47
$C_8H_9O_2Hg$	1.04	0.24	-	-	-	0.64 (HgO)
$C_5H_{13}NCl_2$	1.40	0.75	-	-	0.70	-
$C_8H_5O_3Cl_2$	1.60	0.25	-	-	0.50	-
$C_8H_5O_3Cl_3$	1.39	0.18	-	-	0.65	-
$C_{15}H_{15}O_2NCl$	2.28	0.47	-	0.21	0.19	-
$C_7H_{16}O_2N_3Cl_2$	1.47	0.69	-	0.43	0.26	-
$C_{12}H_{11}O_2N$	2.63	0.49	-	0.15	-	-
$C_{10}H_{12}O_2NCl_2$	2.07	0.51	-	0.14	0.26	-
$C_{16}H_6O_4N$	2.46	0.53	-	0.10	-	-
$C_7H_3O_4N_3S$	1.37	0.12	0.28	0.40	-	-
$C_4H_6N_2S_4Mn$	0.66	0.20	0.97	0.23	-	0.27 (MnO)
$C_6H_{12}N_2S_4$	1.10	0.45	1.07	0.25	-	-
$C_{10}H_6O_2N_2S_2$	1.76	0.22	0.52	0.24	-	-
$C_{12}H_{14}O_6N_2$	1.87	0.45	-	0.12	-	-
$C_{10}H_{12}O_5N_2$	1.83	0.45	-	0.25	-	-
$C_7H_6O_5N_2$	1.56	0.27	-	0.30	-	-
$C_8H_8O_6N_2$	1.54	0.32	-	0.26	-	-
$C_{19}H_{26}O_6N_2$	2.21	0.62	-	0.16	-	-
$C_9H_6O_2SNCl_3$	1.39	0.18	0.26	0.10	0.66	-
$C_{10}H_{19}N_5S$	1.83	0.71	0.26	0.62	-	-

Pesticide	Products of combustion (kg/kg)					
	CO_2	H_2O	SO_2	NO	$CaCl_2$	$Ca_3(PO_4)_2$
$C_7H_{12}N_{15}Cl$	1.53	0.54	-	0.75	0.27	-
Zn_3P_2	-	-	-	-	1.01 (ZnO)	1.37
$C_4H_7O_4Cl_2P$	0.80	0.29	-	-	0.50	0.70
$C_{14}H_8O_2$	2.96	0.35	-	-	-	-
$C_{12}H_{10}O_2$	2.84	0.48	-	-	-	-
$C_{12}H_6O_2Cl_4S$	1.49	0.15	0.18	-	0.62	-
$H_8Cu_4O_7Cl_2$	-	0.16	-	-	0.25	0.18 (Cu)
S_8	-	-	2.00	-	-	-

4.4 Ashes Formed During Incineration of Wastes

There are many problems with the neutralisation of ashes from conventional incinerators, or even more so from hospital-waste incinerators. However, the matter looks very different where wastes are combusted in a cement kiln. For the sake of clarity, we divide the compounds contained in ash into 2 groups:

- the macro-pollutants, i.e. CaO, Fe_2O_3, Al_2O_3 and SiO_2, which enter with the utilised raw materials for the production of clinker and are processed into a final product in the form of clinker.
- from the Table 2 one can see that most metals (not Hg, Tl) are well immobilised into a clinker.

Table 2. Characteristic of retention of selected metals

Metal	Amount introduced to kiln [kg/h	Retention [%]
Cr	0.2558 to 4.6247	99.8553±9.3990
Pb	0.7149 to 19.1765	99.8531±0.2008
Ba	15.00194 to 34.4508	99.8781±9.2339
Cd	0.0392 to 0.1789	≥99.5550±0.4413
As	0.0068 to 2.9891	≥99.8868±0.2916
Be	0.00014 to 0.3253	≥99.8681±1.3278
Se	0.00327 to 0.6082	≥95.4002±1.7779
Ag	0.0199 to 0.2256	99.8420±0.1839
Ni	0.8839 to 2.2794	≥99.9574±0.0108
Sb	0.1602 to 0.3011	≥99.7690±0.0785
Zn	4.1610 to 16.2374	99.7869±0.2598
V	2.2492 to 14.2009	≥99.9922±0.0007

5. SUMMARY

Our research shows that most wastes can be neutralised safely by incineration in cement kiln. Heavy metals are immobilised in a clinker. In the case of combustible wastes, there is a possibility for a partial departure from the use of coal in favour of substitute fuels produced from the wastes. The most favourable situation involves the direct introduction of such wastes into the flame zone in the form of particulates mixed with coal, or else as a liquid pumped in via a lance.

The two remaining methods, i.e. both supply via a sluice and with the aid of a pneumatic discharger, require more expensive containment and are more labour-intensive. They are therefore indicated when it comes to the utilisation of such special wastes as outdated pesticides.

ECOLOGICAL ASPECTS OF CO-INCINERATION OF SOLID WASTE IN CEMENT KILNS

Material flux analysis as a tool for ecological valuation

Uwe Lahl
BZL GmbH, Oyten, Germany

Key words: Co-incineration, cement kiln, material flux analysis (MFA), heavy metals

Abstract: Ecological aspects of co-incineration of solid waste in cement kilns can be analysed by Material Flux Analysis (MFA). MFA is an approved toll for performing an ecological evaluations of intended co-incineration of wastes e.g. in cement kilns, with special regard to emissions of HCl, F, SO_2 and heavy metals. Material flux analysis is also a suitable tool for obtaining information about the resulting quality of the cement itself

1. INTRODUCTION

European Union discussions about co-incineration of solid waste in industrial plants have focused on controversial ecological questions. In Germany, emission standards for waste incineration plants differ considerably from those for other industrial thermal plants like power stations or cement kilns (see Table 1).

Preparing the European Waste Incineration Guideline, this topic was discussed very intensively between the different parties. Following the German waste incineration ordinance (17. BImSchV), so called "mixed-calculation" was adopted to the European rules. But past experience has shown many problems with this "mixing" of different limits.

The draft of the European Waste Incineration Guideline (June 1999) proposes limits for co-incineration that are almost as stringent as the ones for specialised waste incineration plants, as shown in Table 2).

Table 1. Emission limits for different thermal plants in Germany.

daily average values (mg/m³)	power plants (13. BImSchV)	blast furnaces (TA Luft)	cement kilns (TA Luft)	MSWI plants (17. BImSchV)
dust	50	50	50	10
HCl	200	30	30	10
HF	30	5	5	1
SO_2	400	500	400	50
NOx (NO_2)	800	500	1.300 / 500①	200
Cd/Tl	-	-	-	0.05
Hg	-	-	-	0.03
Cd/Tl/Hg	-	0.2	0.2	-
As/Co/Ni/Se/Te	-	1	1	-
Σ other heavy metals	-	5	5	0.5
PCDD/F (ng TE/m³)	-	-	-	0.1

① due to LAI, 1994

Table 2. Limits for co-incineration in the European Waste Incineration Guideline (draft).

daily average values (mg/m³)	MSWI plants (17. BImSchV)	proposed limits for co-incineration in cement kilns from 1.1.2008	proposed exemptions until 1.1.2008 by authorisation
dust	10	30	50
HCl	10	10	-
HF	1	1	-
SO_2	50	50	possible
NOx (NO_2)	200	800	1,200
Cd/Tl	0.05	0.05	-
Hg	0.03	0.05	-
Σ other heavy metals	0.5	0.5	-
PCDD/F (ng TE/m³)	0.1	0.1	-

2. MATERIAL FLUX ANALYSIS (MFA)

Material Flux Analysis[1] is a valuable tool for ecological evaluations. Figure 1 shows the structure of such a calculation process for co-incineration in a cement kiln.

There exist many types of waste, which are regarded as candidates for co-incineration in cement kilns. Table 3 shows a selected list of candidate waste streams.

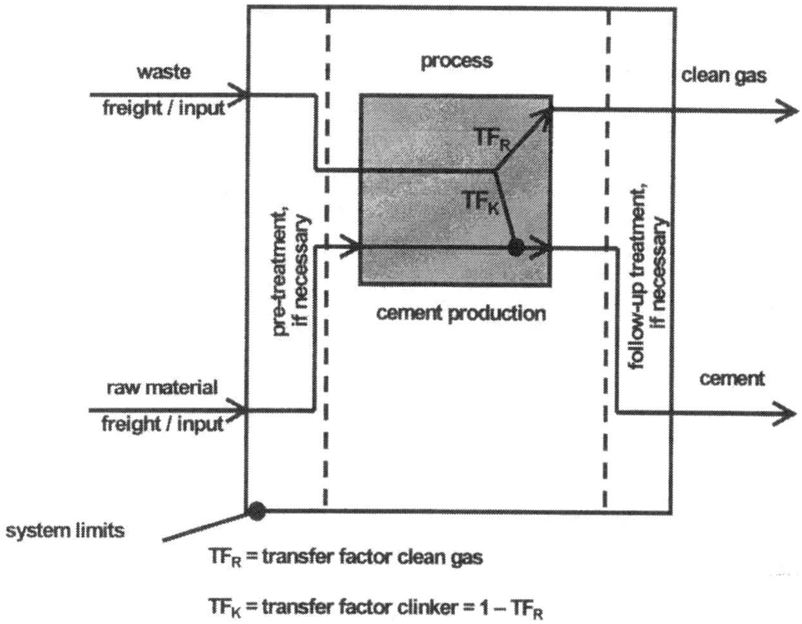

Figure 1. Structure of Material Flow Analysis for co-incineration in a cement kiln – simplified model.

Table 3. Candidates for co-incineration of waste in cement kilns.

LAGA-Code	Waste type
17213	wood with hazardous, mainly organic contaminations
17214	wood with hazardous, mainly inorganic contaminations
18401	rejects from paper and cardboard production/recycling („Spuckstoffe“)
18402	fibre and paper sludges, deinking sludges
18714	packaging with hazardous, mainly organic contaminations
18715	packaging with hazardous, mainly inorganic contaminations
31435	absorbents and filter materials with hazardous contaminations
54209	oil-contaminated resources
54702	oil- und fuel separator contents
55220	solvents, containing halogened compounds
55370	solvents, not containing halogened compounds
55503	varnish and paint waste
57801	light fraction from shredders

Broad and reliable waste characterisation data is required to perform an ecological evaluation by MFA. The following tables show examples of the parameters and data details that are prerequisite to performing MFA evaluations.

Table 4. LAGA code 17213: Wood with hazardous, mainly organic contaminations; n = 186.

Parameter		n	min.	max	average	median	90-quantil
lower calorific value	kJ/kg	29	6,700	30,923	17,227	16,997	21,021
upper calorific value	kJ/kg	13	7,400	20,879	15,448	16,000	19,240
dry substance	%	30	26.3	97.3	85.4	89.5	95.2
water content	%	30	2.7	73.7	14.6	10.6	21.4
glowing loss	%	15	68.4	99.6	87.9	89.4	99.3
glowing rest	%	5	0.24	2.7	1.7	1.8	2.6
Cl	%	41	0.0	13.2	0.5	0.1	0.48
F	%	14	0	0.72	0.06	0.007	0.03
S	%	38	0.01	0.65	0.18	0.13	0.34
N (total)	%	8	0.05	1.5	0.40	0.25	0.87
Sb	mg/kg	14	0.0	20.0	2.6	0.4	5.5
As	mg/kg	26	0.0	234	13.5	1.8	22.5
Pb	mg/kg	27	1.1	315	48.7	18.0	114.5
Cd	mg/kg	30	0.0	5	1	0.41	1.3
Cr (total)	mg/kg	28	0.0	700	80	12	171
Co	mg/kg	11	0.0	44	10.1	5	20.0
Cu	mg/kg	29	1.5	172	39	15	123
Mn	mg/kg	13	20.0	117	67	69	93
Mo	mg/kg	2	0.0	0.0	0.0	0.0	0.0
Ni	mg/kg	21	0.0	60	11	10	20
Hg	mg/kg	30	0.0	4.5	0.9	0.3	2.3
Tl	mg/kg	19	0.0	41.3	6.1	1	28
V	mg/kg	11	0.0	3	1.5	2	3.0
Zn	mg/kg	35	6.3	1,440	209	40	474
Sn	mg/kg	12	0.0	17	7	8	11.4
PCB (Σ 6 congeners)	mg/kg	28	0.0	3,086	315	0.4	1,625
PCP	mg/kg	136	0.0	5,400	323	78.5	805
PCDD/F (I-TE)	µg/kg	1	0.21	0.21	0.21	0.21	0.21

Table 5. LAGA code 18401: Rejects from paper and cardboard production/recycling; n = 98.

Parameter		n	min.	max	average	median	90-quantil
lower calorific value	kJ/kg	59	1,500	27,000	14,423	15,900	21,820
upper calorific value	kJ/kg	59	1,670	30,000	16,024	17,700	24,240
dry substance	%	97	10.9	99.1	57.0	59.6	75.3
water content	%	97	0.9	89.1	43.2	40.4	65.6
glowing loss	%	2	38.7	60.2	49.4	49.4	58.1
glowing rest	%	29	0.9	56.8	9.8	5.1	20.2
Cl	%	73	0.1	5.1	1.1	0.9	2.2
S	%	3	0.1	0.1	0.1	0.1	0.1
N (total)	%	1	5.9	5.9	5.9	5.9	5.9
Sb	mg/kg	3	0.0	0.0	0.0	0.0	0.0
As	mg/kg	7	0.0	2.5	0.4	0.0	1.0
Be	mg/kg	3	0.0	0.0	0.0	0.0	0.0
Pb	mg/kg	10	0.0	93.7	15.9	5.5	30.6
Cd	mg/kg	22	0.0	8.2	1.4	1.0	3.7
Cr (total)	mg/kg	12	0.8	282	29.6	4.6	32.1

Parameter		n	min.	max	average	median	90-quantil
Co	mg/kg	5	0.0	1.8	0.6	0.0	1.6
Cu	mg/kg	12	12.0	55.8	22.2	19.0	30.1
Mn	mg/kg	1	43.8	43.8	43.8	43.8	43.8
Mo	mg/kg	3	0.0	0.7	0.2	0.0	0.5
Ni	mg/kg	6	0.0	9.3	4.9	4.9	8.8
Hg	mg/kg	14	0.0	1.3	0.3	0.0	0.8
Se	mg/kg	5	0.0	0.0	0.0	0.0	0.0
Tl	mg/kg	11	0.0	0.0	0.0	0.0	0.0
V	mg/kg	5	0.0	3.1	1.0	0.0	2.7
Zn	mg/kg	11	74	553	209	170	454
Sn	mg/kg	1	27.9	27.9	27.9	27.9	27.9
PCB (Σ 6 congeners)	mg/kg	13	0.00	0.25	0.07	0.07	0.16

Table 6. LAGA code 57801: light fraction from shredders; n = 245.

Parameter		n	min.	max	average	median	90-quantil
lower calorific value	kJ/kg	19	5,810	20,100	11,426	10,900	18,260
upper calorific value	kJ/kg	13	6,460	22,000	13,780	12,900	20,440
dry substance	%	73	71.7	99.6	93.6	95.8	98.1
water content	%	110	0.4	38.8	7.2	5.6	13.8
glowing loss	%	47	2.8	80.9	36.6	39.7	54.9
glowing rest	%	9	8.3	94.5	49.31	52.2	79.9
Cl	%	25	0.3	18.1	1.9	1.1	2.4
F	%	6	0.0	0.4	0.1	0.0	0.2
S	%	9	0.0	0.5	0.3	0.3	0.5
Sb	mg/kg	5	59.5	560	287	304	465
As	mg/kg	12	5.0	46	16.4	14.8	24
Be	mg/kg	1	0.0	0	0.0	0.0	0
Pb	mg/kg	29	0.5	40,000	5,950	2,900	12,862
Cd	mg/kg	24	9.6	309	65	41.5	99
Cr (total)	mg/kg	23	1.3	2,250	710	434	1,784
Co	mg/kg	2	45.0	52	48.5	48.5	51
Cu	mg/kg	24	2.7	409,800	30,454	8,814	39,100
Mn	mg/kg	5	0.2	893	364	238	717
Mo	mg/kg	3	25.1	30.1	27.5	27.2	30
Ni	mg/kg	23	64.1	3,784	728	446	1,248
Hg	mg/kg	19	0.0	33	6.3	4.0	12.4
Se	mg/kg	3	4.2	8	5.9	5.5	8
Tl	mg/kg	8	0.0	0.8	0.1	0.03	0.3
V	mg/kg	1	27.0	27	27.0	27	27
Zn	mg/kg	25	0.6	19,816	7,081	6,400	12,297
Sn	mg/kg	7	149.5	1600	501	267	1,145.8
PCB (Σ 6 congeners)	mg/kg	213	0.0	296.7	20.9	8.5	49.0
PCDD/F (I-TE)	µg/kg	2	0.094	0.132	0.113	0.113	0.128

Another important prerequisite for performing a MFA evaluation is transfer factors data for the plant. Every plant has its own features and, therefore, its own unique factors. This must be considered while performing the ecological evaluation. Table 7 shows the range of calculated transfer factors obtained by experience or discussions for German cement kilns.

Table 7. Transfer factors for cement kilns[2].

Parameter	Transfer factor into clean gas
Cl	1.30E-01
S	1.00E-02
As	2.00E-04
Cd	5.00E-04
Co	3.00E-04
Cr (total)	4.00E-05
Cu	2.00E-04
Hg	9.30E-01
Ni	7.00E-05
Pb	2.00E-04
Sb	2.00E-04
Sn	3.00E-05
Tl	1.00E-02
Zn	7.00E-05

These factors allow calculating whether a special kind of waste can be incinerated in a defined cement kiln, with exhaust gas stream concentrations maintained within legal limits. Table 8 shows a theoretical example.

Table 8. Demands for an equivalent co-incineration based on the limits for MSWI plants.

Parameter	energy related freight into clean gas
HCl	3.93 mg/MJ
HF	0.39 mg/MJ
SO_2	19.64 mg/MJ
Cd/Tl	0.02 mg/MJ
Hg	0.01 mg/MJ
Σ heavy metals	0.20 mg/MJ

Material flux analysis is also a suitable tool for obtaining information about the quality of the cement itself. Table 9 contains some estimated theoretical figures.

Table 9. Concentration limits for waste derived fuels for 10 resp. 50 % increase of heavy metals in clinker compared to the proposals for limiting waste co-incineration in cement kilns.

Para-meter	Concentration limits for waste derived fuels for 10 % increase of heavy metals in clinker (mg/MJ*)	Concentration limits for waste derived fuels for 50 % increase of heavy metals in clinker (mg/MJ*)	MSW derived fuel[3] (mg/MJ)	Guiding values LAGA[4] (mg/MJ)	Guiding values BUWAL[5] (mg/MJ)
As	1.0	4.8	0.05 – 1.50	1.9	0.6
Be	0.1	0.6	0.005 – 0.01	0.13	0.2
Cd	0.01	0.1	0.1 – 0.3	0.3	0.08
Co	0.4	1.9	0.15 – 0.3	1.2	0.8
Cr	2.5	12.3	3.5 – 10	3.7	4
Cu	0.8	4.1	25 – 50	3.7	4
Mn	14.0	70.2	5 – 25	k.A.	k.A.
Ni	1.6	7.8	1 – 3	3.5	4
Pb	1.2	6.0	7 – 10	10	8
Sb	0.1	0.4	0.4 – 0.6	0.07	0.2
Se	0.1	0.4	0.01 – 0.25	0.2	0.2
Sn	0.3	1.7	0.5 – 1.7	0.4	0.4
Tl	0.1	0.3	0.01	0.15	0.12
V	6.4	32.1	0.2 – 0.5	6.7	4

*) compared to regular raw materials and hard coal; values due to (1), waste derived fuels = 50 % of incineration heat output

REFERENCES

1. Baccini, P., Brunner, P.H., 1991, Metabolism of the Anthroposphere. Springer-Verlag, Heidelberg.
2. Gallenkemper, B., Braungart, M. et al., Untersuchung zur Umwelt- und Gesundheitsverträglichkeit von Substitut-Brennstoffen. Rethmann Entsorgung, Dieselstr. 3, 44805 Bochum
3. BZL GmbH, Oyten/DPU GmbH, Essen, 1999, Beurteilung der Umweltverträglichkeit von thermischen Entsorgungsmaßnahmen. Studie im Auftrag des MURL NRW,.
4. LAGA: Maßstäbe und Kriterien für die energetische Verwertung von Abfällen in Zementwerken. Entwurf, 31.10.1997, zit. in Gallenkemper und Braungart, s.o.
5. BUWAL, o.J., cited by in Gallenkemper and Braungart, s.o.

RECOVERY OF MATERIALS AND ENERGY FROM WASTE IN AUSTRIAN CEMENT WORKS

Karl E. Lorber
University of Leoben, Department of Waste Disposal and Landfill Technology, Peter-Tunner-Strasse 15, 870 Leoben, Austraia

Key words: waste-utilisation, cement-industry

Abstract: During the few last years, the cement industry has developed to an important factor in waste management, as more and more waste fractions and industrial residues are utilised as secondary raw material and secondary fuel in the rotary kiln clinker process. Ideal combustion conditions like high bed-temperatures, oxygen excess and long residence times are creating perfect facilities for the decomposition of organic pollutants. Non volatile inorganic pollutants can be safely fixed in the alkaline matrix of the clinker by counter-current adsorption in the preheater cyclones. In Austria, cement works have become the main disposal facility for used oil, old tyres, fly ash and used foundry sands and recently the sorted plastic fraction of household waste is utilised as a fuel of high calorific value. The fundamental requirements for the thermal utilisation of waste in cement works are no significant increase of the emissions and no deterioration of the cement quality as well as no impairment of work place safety. These conditions only can be met by a packet of control measures, starting with quality control of the input material, development of new emission control techniques (e.g. for NO) for cement plants as well as ambient air quality control programs completed by biomonitoring methods for environmental impact assessment studies. In this paper, the salient features of air quality control measures in modern cement works which are utilising secondary materials are elaborated by way of practically oriented examples.

1. INTRODUCTION

The worldwide cement production in 1991 is reported to reach 1.3 billion tons and for the year 2000 it is estimated to increase to over 1.8 billion tons. Production, handling and use of such quantities of man-made material requires appropriate control strategies and measures to minimize the environment impact effects.

The future development in cement industry is characterized by a sharp increase of capacity for developing countries (i.e. China) as well as a stagnation or decrease of capacity in highly industrialized countries (i.e. Germany). On the other hand, the importance of cement industry as a waste disposal pathway in the latter countries is permanently rising, as more and more residues from industrial processes and from quality controlled refuse collection (RDF) are used as secondary raw materials or secondary fuels in the production of Portland cement. There are three fundamental requirements to be fulfilled, when secondary materials are used for the substitution of conventional raw materials and fuels in cement works:

- The emissions of pollutants must be less or at least not more than before and consequently the ambient air quality is not allowed to get worse.
- The quality of the products (clinker and cement) should be improved or should remain the same but must never deteriorate.
- The State of the Art (standards of technique) and the legal licensing conditions must be met and the production process must not be impaired and the safety of work place must be ensured.

Thus, a packet of measures are necessary in the cement industry, including emission control and safety analyses as well as ambient air quality assurance and environmental impact studies.

2. THERMAL WASTE DISPOSAL IN CEMENT WORKS

Basically, the cement production process provides four opportunities for the thermal utilization of waste and residues, Figure 1:

The introduction of secondary materials can be applied:

- as a raw material (over grinding mill and preheater cyclones or at kiln inlet)
- as a fuel in the secondary firing system for precalcination
- as a fuel in the main firing system for the rotary kiln
- as an additive (or as a grinding aid) in the final cement grinding process.

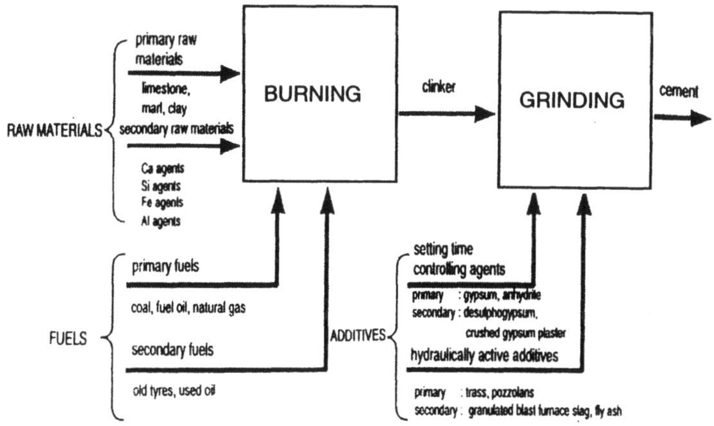

Figure 1. Thermal utilisation of residues and waste in cement industry[1].

Perfect combustion conditions for secondary fuels are available in the main firing system of the rotary kiln, because of the high bed temperature of approximately 1.450°C, the corresponding gas temperature of 2.000°C, the long residence time of about 8 s and more, and the oxygen excess needed in the clinker process.

For secondary fuels, the sorted and shredded plastic fraction of household waste is a most desired material for thermal utilisation, because of the high calorific value (Polypropene, Polyethene: lower calorific value Hu = 44 MJ/kg, PET: Hu = 32 MJ/kg).

However a strict quality control is necessary, as the PVC-fraction should not exceed 2 %, corresponding to about 1 % chlorine in the input material.

At present, used oil, old tyres and processed plastic waste are frequently used as secondary fuel in the Austrian cement industry. For secondary raw material, where Ca, Si, Fe, Al and S additives are required, fly ash, slags, gypsum from flue gas desulfurisation and used sand from foundries are utilised. There are also attempts on pilot-plant scale, to utilise asbestos-cement residues as a Si-additive[2]. The use of secondary materials is beneficial for the cement producers by saving primary fuel and raw material and by obtaining disposal fees for the thermal treatment of waste. The revenue depends on the market situations which can change very fast, depending on offer and demand (e.g. at present, processed plastic waste is scarcely available on the spot-market).

2.1 Input Control Measures

Whenever waste or residues (solid, liquid, gaseous) are utilised in cement works, a very strict input control has to be implemented at the plant. To avoid an increase of emissions of pollutants in the stack-gas and to avoid problems in the clinker process which result in deteriorate quality of the cement, quality control of the input is a must.

Basically, the rotary kiln clinker process offers the opportunity either to decompose organic pollutants at high temperature or to fix inorganic pollutants safely in the solid phase of the clinker. However there are some volatile elements which can cause problems[3]. Mercury e.g., cannot be precipitated or separated from flue gas by the conventional emission control equipment of cement works (preheater cyclones and ESP) and is totally emitted with the stack-gas. Some other less volatile elements like thallium, chlorine and alkali metals do form internal process-cycles by evaporation and condensation, resulting in unwanted enrichment of these elements in the raw meal and raw dust. Provisions have to be made for the partial removal of this elements out of the process by using a bypass-arrangement for the separation of a side stream.

Critical elements in the input material which have to be limited by threshed values are: Hg (< 2 ppm), Tl (< 10 ppm), Cd (< 50 ppm), Cl (< 1%) e.g.

Additionally, there are also limitations for the total amount of waste and residues utilised as secondary materials in the clinker process necessary, which usually are a part of the regulations required for licensing. Generally the maximum massflow [Mg/h] of secondary materials in the clinker process is laid down by the authorities.

3. EMISSIONS AND EMISSION CONTROL

When waste and residues are used as secondary materials in the cement industry, one has to accept that the rotary kiln clinker process will be compared with modern waste incineration technology, especially as far as the emissions are concerned. At present there are controversial discussions going on, whether the same strict emission limits applying to modern waste incinerators should also become compulsory for cement plants.

Waste incineration on one hand and production of cement on the other are different technologies which cannot be compared directly because of significant distinctions, as far as process techniques and process goals are concerned. Traditional cement plants do not have the same highly developed flue gas treatment facilities like modern waste incinerators. But with the exception of NO, CO and VOC (or TOC), due to the specific process conditions a cement plant can meet or even remain under the emission limit values for waste incinerators. Table 1 shows average data for the characteristic emissions of the Austrian cement industry.

Table 1. Characteristic emission data for Austrian cement industry, based on year 1993[4]

Characteristic parameter	Emission Factor [g/t_{CEMENT}]	Stackgas concentrations [mg/m^3], referred to 10 % O_2	Contribution to national emission inventory [%]
Particles (dust)	28.7	16.8	-
Nitrogen oxides (NO_2)	1,296	739	3.23
Sulfurdi oxide (SO_2)	191	98	1.36
Inorganic chlorine (HCl)	4	2.2	-
Inorganic fluorine (HF)	0.37	0.23	-
Total organic carbon (TOC)	49.5	28	0.06
Carbon monoxide (CO)	2,209	1,348	0.81
Carbon dioxide (O_2)	632,000	371,600	5.41
pyrogene CO_2	212,400	125,600	1.83
calcination CO_2	419,600	245,900	3.58
Cadmium (Cd)	0.011	0.0046	0.53
Thallium (Tl)	0.0086	0.0047	-
Arsenic (As)	0.0047	0.0028	0,24
Cobalt (Co)	0.0338	0.0200	-
Nickel (Ni)	0.0028	0.0017	0.04
Lead (Pb)	0.0153	0.0090	0.07
Mercury (Hg)	0.0132	0.0066	1.47
Chromium (Cr)	0.0064	0.0038	0.43
Zinc (Zn)	0.1186	0.0700	0.55
Dioxine (PCCD/F)	< 0.17 mg/t	< 0.1 ng/m^3	-

For conversion of the Emission Factor [g/t_{CEMENT}] to [g/$t_{CLINKER}$], the figures show in Table 1 have to be multiplied by the factor of 1.32. Principally two different operation conditions have to be considered for the clinker-process: direct operation and interconnected operation, where the raw gas stream passes the grinding mill. The emissions referred to 10 % oxygen in the stackgas are the same, except for SO_2 which is significantly higher for the direct operation mode.

3.1 Emission Control

As shown in Table 1, CO and NO are the main emission problems created during the clinker-process. In the cement industry, a high CO-concentration in the stack-gas does not indicate incomplete combustion in the rotary kiln, as the CO emitted is mainly formed out of the carbon content of (also natural) raw materials in the preheater cyclones. For safety reasons, the ESP is shut down automatically whenever a certain threshold limit for CO is exceeded, leading to short periods (about 30 seconds only) of uncontrolled dust emissions. So far, this problem has not been solved satisfactory. The specific NO-emission problem in the cement industry is mainly caused by thermal NO-formation at high temperatures of the clinker process.

Different approaches have been tried to reduce the NO-emission, like:

- Staged secondary firing technology and low NO_x-burner for primary firing systems
- Feeding of shredded old rubber tyres at the kiln inlet: $2\ CO + 2\ NO \rightarrow 2\ CO_2 + N_2$
- Selective non-catalytic reduction of NO with urea or ammonia at 900 ° - 1,000 °C at the kiln inlet side (SNCR-process), Figure 2,
- Selective catalytic reduction of NO (SCR-process: $NO_2 + NO + 2\ NH_3 \rightarrow 2\ N_2 + 3\ H_2O$).

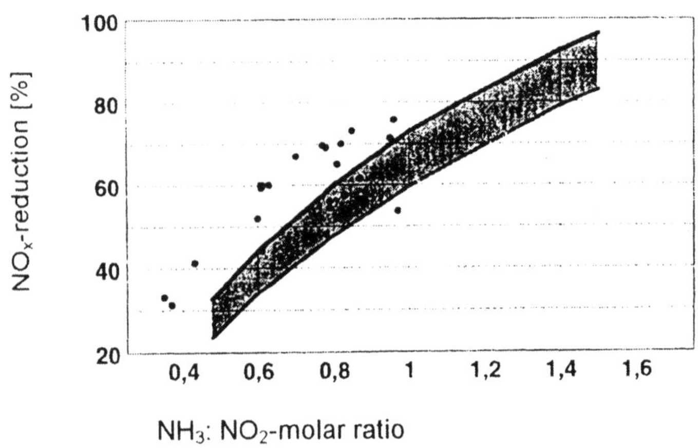

Figure 2. Efficiency of the ammonia-water injection for NO_x-reduction as a function of the NH_3/NO_2 molar ratio[5].

Since 1995, a "high-dust" SCR-pilot plant is operated at Tavernola, Italy. The SCR-catalyst is placed in the raw gas stream right before the roller grinding mill at 280 - 320 °C. So far only positive results are reported with NO_x-reduction rates up to 90 %[6].

Heavy metals (which are not too much volatile like mercury) and acid gases (HCl, HF and SO_2) are efficiently removed out of the raw gas stream by counter-current adsorption on alkaline raw meal in the preheater cyclones. The conventional flue gas cleaning system in cement works also contains an ESP for the separation of dust.

4. AMBIENT AIR QUALITY CONTROL MEASURES

Whenever a new cement plant gets commissioned in Austria or before a new type of secondary material is utilised in an already licenced plant, a prognosis of the expected emissions has to be submitted to the authorities. As long as the threshold values laid down for the input material are met, usually no increase of the already existing emissions can be expected, as the main pollutants CO, VOC (TOC) and NO are mainly depending on the rotary kiln clinker process and only to a less extend on the input material. Nevertheless, based on the estimated emissions, an environmental impact assessment study has to be accomplished. For that purpose, relevant meteorological data like wind direction, wind speed, temperature and humidity of ambient air e.g. are collected in the region for a minimum period of one year with the help of mobile or stationary ambient air monitoring stations. Additional parameters measured are dust (Bergerhof-methode), NO, NO_2, SO_2 and CO. Using the calculated emission data and the meteorological parameter collected, dispersion modelling is done for normal and "worst case" conditions. Worst case conditions are assumed when the wind speed is less than 0.8 m/s together with most disadvantageous inversion layers above the stack in an orographically difficult (hills, impactation slopes) area. The results are maximum ambient air concentrations, which are used for the prediction of health effects.

For this purpose, the socalled "Unit-Risk" - model of EPA is used, where "Virtually Safe Dose" as well as "Acceptable Dose" regions are defined for toxic pollutants.

Additional, a socalled "safety-risk and emergency situation analysis" is carried out, assuming "worst case" conditions for the workers and using the IDHL (immediately dangerous to live and health) - list of EPA for risk assessments.

Finally, also a biomonitoring program is installed and samples (gras, soil 0 - 10 cm, leaves and needles from trees) are investigated for S, Cl, F and heavy metals. In areas with predicted maximum concentration of pollutants (e.g. at the expected impactation points), standardised indicator plants, like "Lolium multiforum ssp. Lam.", type "Lema", are exposed 1.5 m above ground level for 28 days each and analysed for Cd, Tl, As and Hg, Table 2. The results are compared with the calculated deposition - rate [$mg/m^2 \cdot d$] of the emitted dust particles, assuming an average sedimentation velocity of 0,035 m/s.

The ambient air quality control program mentioned above usually requires two years, one year before and one year after the plant starts with operation. This enables the documentation of evidence, when the former situation is compared with the actual one.

Table 2. Standardised biomonitoring with indicator plant "Lolium multiforum ssp. Lam". Scale for evaluation of environmental impact effects by sedimentation of dust.

concentration found in exposed indicator plant [mg/kg DM]			
Element analysed	Natural background	Low environmental impact	Significant environmental impact
Cadmium (Cd)	< 0.25	0.25 - 0.4	> 0.4
Thallium (Tl)	< 0.1	0.1 - 0.25	> 0.25
Mercury (Hg)	< 0.03	0.03 - 0.06	> 0.06
Arsenic (As)	< 0.25	0.25 - 1.0	> 1.0

DM = dry material

5. CONCLUSION

Two goals can be obtained at the same time, when waste and residues from industry is used either as secondary raw material or as secondary fuel in cement works:

- Safe disposal of waste and unwanted residues by thermal treatment.
- Saving of energy and primary raw material followed by improved economic production conditions in cement industry.

To guarantee no impairment of the emission situation and not deterioration of the cement quality, the implementation of a stringent input control program is a must. Therefore the cement works should have a well-equipped analytical laboratory.

When the rotary kiln clinker process is compared with modern waste incineration technology, one has to consider that the low emission limits for CO, VOC (TOC) and NO up to now cannot be met in cement industry, because there are no effective pollution abatement systems available for these compounds. For NO, interesting results are reported from a "high dust" SCR-pilot plant, reaching 90 % NO_x-reduction..

REFERENCES

1. Liebl, P., and Gerger W., 1993, Benefits and limitations when using secondary materials, *Zements-Kalk-Gips* (ZKG) International No. 10/93.
2. Liebl, P., and Gerger W., 1993, Benefits and limitations when using secondary materials, *Zements-Kalk-Gips* (ZKG) International No. 10/93.
3. Sprung, S., Hoenig V., and Schneider, C., 1997, New process for utilising asbestos cement products, *Zement-Kalk-Gips* (ZKG) International No. 1/97.

4. Kirchartz, B., 1994, Reaktion und Abscheidung von Spurenelementen beim Brennen des Zementklinkers, *Schriftenreihe der Zementindustrie*, Heft 56/1994, Beton-Verlag GmbH, Düsseldorf, ISBN 3-7640-0333-2.
5. Hackl, A., and Mauschitz, G., 1995, Emissionen aus Anlagen der österreichischen Zementindustrie, editor: Zement + Beton Handels- und Webeges.m.b.H., Reinserstraße 53, A-1030 Wien.
6. Rose, D., and Brentrup, L., 1995, Effective emission reduction when using secondary materials at the Siggenthal cement works in Switzerland, *Zement-Kalk-Gips* (ZKG) International No. 4/1995.
7. Leibacher, U., 1996, NO_x elimination by the SCR-process for cement kiln systems, *Zement-Kalk-Gips* (ZKG) International No. 6/94.

PCDF/D EMISSION FROM WASTE UTILISATION IN CEMENT KILNS

Marzenna R. Dudzińska and Zdzisław Kozak
Instytute of Environmental Protection Engineering, Lublin University of Technology, 40, Nadbystrzycka Str., 20-618 Lublin, Poland

Key words: waste incineration, PCDF/D emission, clinker production,

Abstract: Thermal waste utilisation is strongly debated environmental issue, especially siting a new incineration plant. An alternative possibility is to utilise existing industrial installations for waste incineration, what is strongly support by directives of European Community. This paper gathers the current state of knowledge on hazardous emissions such as PCDFs and PCDDs from waste incineration in cement kilns. Although, condition of clinker production are hardly preferable for PCDF/Ds. formation, some mechanisms allowing PCDF/D synthesis are discussed as well as mechanisms inhibiting formation of PCDF/Ds.

1. INTRODUCTION

One of the most important environmental problems of new millennium is growing amount of solid and liquid wastes.

Generally, the methods applied for solid waste utilisation can be disposal in landfills, composting, incineration and recycling. Placement of waste in the landfills is economically attractive, storing materials which cannot be reused is not economically reasonable.

Recycling appears to be the best solution, as it may be economically efficient and environmentally friendly. But unfortunately, not all materials can be recycled and due to sanitary requirements recycled products cannot always be used. Composting can also be considered a form of the recycling, but it is limited only to biodegradable wastes (biomass recycling), and must be free of heavy metals. If recycled wastes contain hazardous contaminants, such as heavy metals or non-degradable organics, these contaminants are only recycled, not treated or destroyed.

Thermal waste utilisation is strongly debated environmental issue. Incineration is a controversial topic for environmentalists and policy makers, since it usually meets opposition from local communities, especially when local siting of a new incineration plant is proposed.

Although incineration is called "total destruction" it normally reduces the volume of waste only up to 80%. It also results in emission of fly ash, acidic gases and heavy metals, as well as hazardous volatile organics. Devices to abate emissions generate high amounts of liquid hazardous wastes and are expensive to operate. The most controversial emissions are the emissions of polychlorinated dibenzo-p-dioxin and dibenzo-p-furans (PCDF/Ds). Through controversial, incineration is still a variable solution and, taking into consideration all advantages and disadvantages, there is no other alternative for addressing the growing amounts of wastes.

Thermal waste utilisation should be considered a recycling method since energy is recovered in the process. It is both a valuable methods for destroying organic, non-degradable substances and a method of recycling since it the waste burned can serve as a substitute for conventional fuel. The amount of pollutants emitted in incineration processes depends on the composition of the input and the efficiency of the incineration process. All solutions and options that promote the minimisation of both solid by products and gaseous emissions should be considered.

One possibility is to utilise existing industrial installations for waste incineration. Such solutions are strongly supported by directives of European Community. Optimal conditions, as recommended by the directive CEE/COM(93)296, are presented in Table 1. One type of facility meeting the qualifications for waste incineration is a rotary cement kiln for clinker production.

Table 1. Optimal parameters for waste incineration according to the EU directive CEE/COM(93)296.

Temperature	For 'organic' Cl content	
	- below 1%	1123 K
	- above 1%	1473 K
Oxygen demand	Above 6%	
Reaction time	More than 2 s	
CO content	In 100% of day average	< 50 mg/Nm^3
	In 95% of 10 min. average	< 150 mg/Nm^3

2. PCDF/D FORMATION IN CHEMICAL AND THERMAL PROCESSES

Polychlorinated dibenzodioxins (PCDDs) and dibenzofurans (PCDFs), in contrast to other chlorinated chemicals, have never been commercially manufactured or found to be of any benefit or known use. They enter the environment not as primary products, but as unintentional trace impurities. PCDFs and PCDDs are considered to be very toxic and hazardous substances. Because of the different numbers of chlorine atoms in the molecules, 75 polychlorodibenzo-p-dioxins and 135 polychlorodibenzo-p-furans might be formed. Almost all of them have been found in the environmental matrices (air, soil, sediments and animals tissue)[1], but only the 17 congeners that are considered to be of the highest toxicity are usually measured.

The PCDF/Ds level is presented as TEQ, e.g. the sum of the 17 congeners levels multiplied by their toxicity factors (TEF), and only this value is regulated by law in many countries[2]. The highest toxicity factor TEF = 1 was attributed to 2,3,7,8-tetrachlorodibenzo-p-dioxin (2,3,7,8,-TCDD). In some animals, 2,3,7,8-TCDD exhibited extreme toxicity, higher than any other synthetically molecule. It became to be recognised as the most hazardous substance for the environment and human beings. Furthermore, all 2,3,7,8-substituted compounds persist in the environment and have been found to strongly bioaccumulate in fish and mammals. In most countries, the PCDF/D level in gaseous emissions is limited (TEQ = 0,1 ng/ Nm^3, in the EU countries). Daily uptake, recommended by WHO should not exceed 10 pg/kg of body mass, although in a number of countries lower limits are permitted. Measurement and calculation of TEQ is considered to be sufficient for estimation of PCDF/Ds hazard to the humans and environment.

The primary sources of PCDFs and PCDDs in the environment can be divided into four categories: chemical reactions, thermal reactions, photochemical reactions and enzymatic reactions[3].

Chemical reactions, where PCDF/Ds are formed from their precursors in industrial syntheses, resulted in unintentional contamination of products like chlorophenols, chlorophenoxy herbicides, pesticides and PCBs[4]. Production and use of these chemicals are currently banned or strictly regulated, but in 1960s and 1970s, these were widely used products. As a result, they were a major source of PCDF/Ds in the environment. Other PCDF/Ds sources are paper pulp bleaching, dry cleaning distillation residues and chlorine gas production using graphite electrodes.

Thermal reactions contributing to the PCDF/D emissions are associated with technological and domestic combustion. PCDF/Ds have been found in the emissions of various combustion processes independent from the fuel source: municipal and hazardous waste incinerators, power plants with fossil fuels, automobile exhaust, private heating and fire places, wood and forest fires, cigarette smoking as well as accidents like PVC or PCB fires[5]. Copper smelters and steel production are also thermal sources[6], but it is mainly waste incineration that raises public concern.

Photochemical reactions can result in the formation, as well as the degradation, of PCDF/Ds. These reactions are being carefully examined, as most combustion and incineration sources produce emissions directly into the atmosphere and they undergo long distance transport.

According to the recent evidence, enzymatic reactions leading to PCDF/D formation might occur under natural environmental conditions in sewage sludges and during composting processes[7]. The latest research, which identified PCDF/Ds in composting processes, created additional problems for the waste compost market.

PCDD/Fs are emitted or spilled from many processes and background levels of PCDF/Ds have also been reported not only in a series of abiotic reservoirs like soil and sediments, but also in air, water and snow. But, according to Rappe[3], patterns found in environmental samples are typical combustion patterns, indicating that various combustion processes are the major ultimate source for this worldwide contamination. PCDD/F at detection levels of GC/MS/MS were found in the 6000 years old layers in Japan, as well as in 13th century layers in Northern Poland. This suggests that trace levels of PCDD/F can be attributed to wood combustion and activities of early man.

The public should be made aware of all processes that produce "dioxin" to limit its formation, but the presence of PCDDs and PCDFs in flue gases and fly ash from incineration processes is already universal and can be attributed to three possibilities:

1. PCDF/Ds are already present in the trace amounts within the fuel and are not destroyed by the combustion process,
2. PCDD/Fs are formed during the process from precursors,
3. PCDF/Ds are produced in '*de novo*' synthesis of other, non-chlorinated organic substances.

Altwicker[8] proposed four mechanisms for PCDF/Ds formation in thermal processes:

- from carbon matrices on the particle surfaces ('*de novo*' synthesis),
- from structurally similar precursors,
- from simple compounds containing C, Cl, H, O,
- from unstable transition products of combustion.

Many parameters may effect the reduction or increasing in PCDF/D emission level, but not all of them have been fully recognized by the scientific community. These usually include temperature, oxygen level, contact time for particles and precursors, presence of chlorine and its compounds, catalytic effect of dust particles, catalytic effect of some metals, mainly copper, and inhibiting effect of sulfur dioxide. The oxygen level, as well as the presence of acidic gases, may also influence PCDF/D formation. Some of these parameters are attributed to the characteristic of waste being combusted, others to the technological conditions of the combustion process.

3. CLINKER PRODUCTION PARAMETERS INFLUENCE ON PCDF/D EMISSION

It is generally believed that PCDF/Ds are formed in combustion processes within the temperature range 200 - 850°C, but the optimal range is 300-325°C[9]. In the installations where plastic-containing electronic equipment is incinerated, quick cooling was implemented to omit the temperature range favourable for "dioxin" formation[10]. Currently, high temperature incineration processes are applied, with temperatures high enough to avoid PCDF/Ds synthesis. According to Table 1 and the EU directive, temperatures above 850°C are recommended, and for wastes containing more then 1% of chloride, temperatures greater than 1200°C are recommended. The EU directive also recommends a minimum contact time of 2 secs for fumes within the high temperature zone. The reaction time is very important because it has to be sufficient for the total destruction of even stable organics, but also because; it may have impact on the recombination reactions of the intermediate products into "dioxins" in heterogenic catalytic processes.

The production of clinker is a high temperature process, with temperatures above 1400 – 1500°C in the calcination zone. Large amounts of the raw material used in the process causes temperature stability conditions in cement kilns. So, generally the technological parameters are not preferable for PCDF/Ds formation.

It was found that the oxygen level plays an important role in '*de novo*' synthesis. "Dioxin" emissions increase with increase in oxygen concentration[11,12]. However, insufficient oxygen results in incomplete combustion and the formation of soot particles that are discharged as emissions. The increasing emissions of PCDF/Ds can be related to oxygen levels catalytic processes A minimum oxygen level of 6% is recommended to limit PCDF/D formation. For cement kiln incineration, the level usually must be raised to greater than 6%, based on the technological requirements of the process.

The catalytic effect of soot particles on "dioxin" emissions that are present in the fumes as a result of incomplete incineration is the most common effect described in the literature. The contact surface plays an important role in the *'de novo'* synthesis[13, 14], as well as in the synthesis from the radicals[15,16].

In the *'de novo'* synthesis, the soot particles are the precursors, and their surface may be a heterogenic contact surface for chlorine (from HCl), water particles and oxygen, according the scheme:

$$C + HCl + O_2 + H_2O \rightarrow PCDF/PCDD$$

During the clinker manufacturing, the soot is incorporated into the cement structure. There is no information about the catalytic influence of the surface of cement dust particles. However, there is evidence that HCl has a role in the competitive reaction with calcium oxide:

$$CaO + 2HCl \rightarrow CaCl_2 + H_2O$$

The generated calcium chloride is retained at the electrostatic precipitator.

4. WASTE COMPOSITION INFLUENCE ON PCDF/D FORMATION

Chlorine atom presence is an obvious condition for polychlorinated dibenzodioxins (PCDDs) and polychlorinated dibenzofurans (PCDFs) formation. However, literature data does not show the clear relationship between chorine input and "dioxin" emission level. For many years, the presence of polyvinyl chloride (PVC) in domestic wastes was recognised as a precursor of PCDF/D formation in municipal waste incinerators[17]. However, it has not been confirmed by the emission measurement for highly chlorinated waste combustion processes[18].

It appears that the relationship between the chlorine level within a waste and the chlorine level in the emission is not a simple one. Nevertheless, input of chlorine-containing materials into municipal waste incinerators is strictly limited and regulated in many countries.

The chorine in the processes of waste incineration in cement kilns may originate from either organic waste substituting fuel and introduced directly into the high temperature kiln zone or inorganic wastes including metal chlorides, substituting the raw material. But because of technological rigor and the prime goal - the high quality clinker - wastes incinerated cannot contain too high level of chloride and too high level of metals, especially heavy ones. Both parameters should be controlled and limited.

The incineration of chlorine-containing organics waste can be describe by the overall reaction:

$$C_xH_yO_zCl_n + O_2 \rightarrow CO_2 + H_2O + HCl$$

Hydrochloric acid formed in this reaction can react with calcium oxide present in the raw material to generate calcium chloride, which can be retained at the electrostatic precipitator.

The increased level of "dioxin" emissions from copper smelters shows the possibility of catalytic effect of some metals, especially copper, on the PCDD/F formation[6]. Huang and Buekens[9] claim that copper(II) ions act as catalyst in the '*de novo*' synthesis. To a lesser extent, similar effects was observed in the presence of Fe^{3+} , Pb^{2+} and Zn^{2+} ions. No influence of Sn^{2+}, Fe^{2+}, Mn^{2+}, Ni^{2+} and Cd^{2+} ions was found.

In recent years, the inhibiting effect of SO_2 was confirmed by several scientists, although the mechanism still is not fully understood. According to Huang and Buekens[19] sulphur dioxide may mask or screen the active centers of particle surfaces, which usually act as catalysts. Reduced emission of PCDF/D in the presence of SO_2 was confirmed by Lindbauer[20] and Ogawa et al.[21]. Ogawa[21] proposed the following reaction mechanism of Cl_2 to HCl:

$$SO_2 + Cl_2 + H_2O \rightarrow SO_3 + 2HCl$$

which may cause reduction in PCDF/D formation.

The emission of acidic gases is not a problem of incineration processes in cement industry because of the alkaline environment inside cement kilns. This presents a real advantage of cement kilns over conventional waste incinerators. Sulfur dioxide, as well as HCl, can be neutralised in a reaction with calcium oxide:

$$CaO + SO_s \rightarrow CaSO_3$$

and generated $CaSO_3$, which could be retained at the electrostatic precipitator, similarly to that previously described for $CaCl_2$. It is difficult to say if SO_2 from the sulfated coal would influence processes in the alkaline environment of cement kiln. It seems more probable that HCl neutralisation may cause the reduction of the PCDD/F emission.

5. LEVELS OF PCDF/D EMISSION FROM WASTE COMBUSTION IN CEMENT KILNS

As discussed earlier, all incineration processes, independent of fuel source, generate PCDF/Ds pure coal combustion is also a source of emission. If the temperature in a rotary cement kiln exceeds 1470°C for more then 3 s, the conditions are not preferable for dioxins formation. In Polish plants, high sulfur coal is usually used as a fuel. The natural occurrence of SO_2 in the process might suggest another disturbance in PCDF/D formation. However, coal contains traces of chlorine.

Emission measurement from cement plant that utilises coal as fuel shows PCDD/F at the detectable level, but below the allowable standards limit.

This was confirmed for Rejowiec Cement Plant, where such measurements were performed by Dudzińska *et. al.*[22] (Project PBZ 20-04/97) and the emission of 0.0515 ng/Nm^3 (I-TEQ) was found.

Different situations developed when wastes are incinerated in industrial facilities. In Poland, attempts were made to incinerate different types of wastes, including hazardous ones, in different facilities. Power Plants are an example. As it was found by Grochowalski and Wybraniec[23] for Tarnów Power Plant, where test utilisation of vinyl chloride was undertaken in November 1994, most of PCDF/D emission is attributed to fly ash rather then flue gases. This, on one hand, confirms the mechanism of *'de novo'* synthesis, catalysed by soot particles. On other hand, it causes the PCDF/D contamination in the fly ash and limits its future use. The highest emission levels found by Grochowalski and Wybraniec[23] were 0.051ng/Nm^3 for flue gas and 92 ng/kg for fly ash (as I-TEQ).

PCDF/D formation mechanism for cement kiln waste incineration processes has not been discussed in the literature as often as processes for regular waste incineration plants. Based on available data, it can be stated that partial substitution of conventional fuel by organic wastes does not increase significantly PCDF/D emissions, compared to emissions from the kiln run using 100% conventional fuel. Any exceedances of allowable limits were not noticed in the USA, but federal standards allow emissions up to 0.5 ng/m^3. Results from selected Environmental Protection Agency (EPA) reports[24,25] are shown in Table 2. Fuel substitution measurements were made using organic solid and liquid wastes only.

Table 2. Emission of PCDF and PCDD, (as I-TEQ) from the cement kilns (*compiled literature data*[24, 25]).

Fuel	I-TEQ, ng/Nm^3
100% coal	0.076
Coal + liquid and solid wastes	0.097
Liquid + solid wastes (100% waste fuel)	0.125

In Polish cement plants, various attempts were made to utilise either organic wastes as a substitute for fuel, or inorganic wastes as a substitute for raw material. An example is the study conducted in Rejowiec Cement Plant. Inorganic wastes came from the electroplating plant and contains metals, such as Zn, Cu, Pb, Cd, Ni, and Fe in the form of oxides and hydroxides. Organic wastes incinerated in cement kilns did not contain copper. Inorganic wastes contained 0.08% copper.

Table 3. Emission of PCDF and PCDD, (as I-TEQ)

	I-TEQ, ng/Nm^3
Coal + 10% inorganic wastes	0.082
Coal + 10% organic wastes	0.069

An estimation of the effect of the discussed above parameters on "dioxin" formation was difficult to obtain because of the limited ability to obtain measurements. Generally, the addition of organic wastes with partly substituted fuel was expected to result in a smaller increase in PCDF/D emissions than the addition of inorganic wastes as a substitute for raw material, although, organic wastes contained chlorine while inorganic did not. Inorganic wastes contained metals in the form of oxides and hydroxides, but chlorine traces were still present.

A study of the effects of waste co-incineration in cement kilns on the environment would be an important research topic. Although the increase of PCDF/D levels from waste incineration in cement kilns compared to conventional processes is expected, the increase should not exceed the allowable limit. Any PCDF/D attributed to dust particles should not cause environmental problems, as these particles are recycled into product in the Cement Plant. PCDF/Ds are very persistent in soils and fly ash, and no transport under normal environment conditions was found[26]. Persistence in the cement was not studied but might be expected.

6. CONCLUSION

Available research data do not confirm exceeding of the 'dioxin' emission from clinker production using organic wastes as a fuel substitute. Our study suggests that waste incineration in clinker rotary kilns does not cause any additional negative impacts to the environment connected with PCDF/Ds emissions. In some cases, for selected wastes of low chlorine content, emissions were even less than emissions from total coal combustion. When raw material is partially substituted by inorganic wastes, containing chlorine, the situation is more complicated, but most of the "dioxins" emitted in the process, appears to be fly ash PCDF/Ds which are captured on precipitators. Some of PCDF/D may be destroyed in the high temperature process.

PCDF/Ds are extremely persistent in the environment. Therefore, any means of possibly reducing their level should be treated as being beneficial. The benefits should promote additional support for waste incineration in cement kilns for minimization of the 'dioxin' problem.

REFERENCES

1. Hagenmaier, H., Linding C., and She, J., 1994, *Chemosphere* 29: 2163-2174.
2. Safe, S., 1992, *Chemosphere* 25: 61-64.
3. Rappe, C., 1994, *Fresenius J. Anal. Chem.* 348: 63-75.
4. Heindl, A., and Hutzinger, O., 1986, *Chemosphere* 15: 2001- 2008.
5. Hutzinger, O., and Fiedler, H., 1989, *Chemosphere* 18: 23- 32.
6. Karasek, F.W., Viau, A.L., and Gonnard, M.F., 1983, *J.Chromatography* 270: 227-231.
7. Oberg L.G., Anderson R., and Rappe C., 1992, *Organohalogen Compounds* 9: 351-354.
8. Altwicker E.R., 1996, *Journal of Hazardous Material.* 47: 137-161.
9. Huang, H, and Buekens, A., 1996, *The Science of Total Environment* 193: 121-141.
10. Szpadt, R., 1992, Utylizacja i unieszkodliwianie złomu elektronicznego i elektrotechnicznego, in: *Mat. Konf., "Komputer W Ochronie Środowiska"*, Poznań, pp. 81-92.
11. Vogg, H., Metzer, M., and Stieglitz, L., 1987, *Waste Management Res.* 5: 285-292.
12. Addink, R., and Olie, K., 1995, *Environ. Sci. Technol.* 29: 1586-1591.
13. Vogg, H., and Stieglitz, L, 1986: *Chemosphere* 15: 1373-1376.
14. Altwicker, E.R., Konduri R., Lin, C., and Milligan, M.S., 1992, *Chemosphere* 25: 1935-1944.
15. Olie, K., Vermeulen, P., and Hutzinger, O., 1977, *Chemosphere* 6: 455-459.
16. Konduri,R., and Altwicker, E., 1994, *Chemosphere* 28: 23-45.
17. Wilson, D.G. (ed.) 1997, *Handbook of Solid Waste Management*, Van Nostrand Reinhold, New York.
18. Hawley-Fedder, R.A., Parsons, M.L. and Karasek, F.W., 1984, *J. Chromatogr.* 315: 211.
19. Huang, H., and Buekens, A., 1995, *Chemosphere* 31: 4009-4014.
20. Lindbauer, R.L., 1992, *Chemosphere* 25: 1409-1414.
21. Ogawa, H., Orita, N., Horaguchi, M., Suzuki, T., Okada, M., and Yasuda, S., 1996, *Chemosphere* 32: 151-157.
22. Dudzińska M.R., Kozak Z, and Pawłowki L, 1998, An Attempt to estimate the PCDF/D Emissions from Cement Kilns in: *Chemistry for the Protection of the Environment 3, Environmetal Science Research Vol. 55.* Plenum Publishing, pp. 173-179.
23. Grochowalski A., and Wybraniec, S., 1996, *Chem. Anal. (Warsaw)*, 41: 27-31.
24. Brandsome, R., and Mourinham, R.E., 1995, *EPA 600/2-87-095.* U.S. Environmental Protection Agency, Office of Research and Development, Hazardous Waste Engineering Research Laboratory, Cincinnati, OH, 67 pp.
25. *Candance Environmental Energy Inc. Report*, 1995, presented by M.R.Benoit at Seminar in Kujawy Cement Plant, Poland, April 1995.

26. Schramm, K.W., Merk, M., Henkelmann, B. and Kettrup, A., 1995, *Chemosphere* 30: 2249-2257.

UTILISATION OF SEWAGE SLUDGE IN CEMENT KILNS

[1]Marek Kotowski, [1]Lucjan Pawłowski, [1]Aneta Czechowska and [2]Urszula Kotowska

[1]Institute of Environmental Protection Engineering, Lublin University of Technology, 40, Nadbystrzycka Street, 20-618 Lublin, Poland, [2]Institute of Agrophysics, Polish Academy of Science, 4, Doświadczalna Str., 20-290 Lublin, Poland

Key words: sewage sludge, incineration, cement kiln, utilisation

Abstract: Research results of mechanically-dehydrated sewage sludge utilisation in cement kiln are presented. The fuel mix was obtained by mixing sewage sludge with fuel oil and waste oil. Non-ionic surfactants Rokafenol N8 and Rokacet S7 were used as emulsifier. Technical tests of utilisation of sewage sludge (supplied to a cement kiln with clinker slurry) and combustion of its mixtures with fuel oil were run at the Rejowiec Cement Works. It was found that the utilisation of sewage sludge in cement kilns does not cause any excess harassment to the environment.

1. INTRODUCTION

The choice of methods by which to dispose of or utilise the sludge generated at municipal wastewater treatment plants is dependent on the chemical composition and on such physico-chemical parameters of the sediments involved[1] as contents of heavy metals, organic and mineral substances and degree of hydration. The most favourable situation is obviously that in which disposal (utilisation) is combined with the use of valuable components of sewage sludge as in:

- use for non-industrial purposes, and hence in manuring, soil improvement and reclamation, etc.[2];

- the solidification of sludge with a view to obtaining materials that can be applied in roadbuilding, the production of sorbents, etc.[3];
- energetic use through the processes of thermal combustion or gasification[4,5].

Each of the methods referred to has its advantages and disadvantages. The most radical method is burning, with the basis of the process being the maximal use of heat from it. At present the most frequently-applied means of thermally disposing of wastes are those involving incineration in rotary drum furnaces, in fluidised-bed boilers and in chamber furnaces.

Among the newest means by which to make energetic use of sewage sludge are: the Bojner system from AB Torkapparatet and the KREPRO system from the Swedish firm Kemira[6]. These are thermal methods whose application allows products capable of being used in nature to be obtained.

When various reasons rule out the natural methods of using sewage sludge or the methods involving solidification, while the use of more sophisticated technologies in the energetic use of wastes is unrealistic, total disposal may be achieved through the burning of sewage sludge in a cement kiln. In this case, the chances of using installations protecting the environment will be of decisive importance.

The aim of this study is to develop a method by which to generate a constant mixture of mechanically-dehydrated sewage sludge with fuel oil and to determine the possibilities for the fuel so-obtained to be used in the firing of a cement kiln. The mineral substances contained in sewage sludge should be bound up in clinker to a very great extent, such that there is almost total and permanent immobilisation[7-10]. The components involved are Ca, Fe, Si, K, P, Al and heavy metals. During the combustion process, sulphur may be liberated into the gaseous phase and pass to the atmosphere with waste gases. Its dry-mass content in sewage sludge is considerable and comparable with that in hard coal other than of the highest quality. The problem of the liberation of sulphur as sludge is burnt should therefore be taken very seriously.

2. METHODS

In order to attain the objectives set, detailed research on the chemical composition of sewage sludge from the Hajdów (Lublin-Poland) Wastewater Treatment Plant was carried out in the period July 1998 to March 1999. Analyses concerned the content of water, organic substances, sulphur and metals, with the latter including the heavy metals Zn, Cu, Cr, Pb and Ni. The heat of combustion of the sludge was determined, and an energetic balance drawn up for the processes of combustion of both wet and dry sludge.

The fuel mix was obtained by mixing the sewage sludge with fuel oil and waste oil generated by the rolling mills at the Sendzimir Steelworks in Nowa Huta. In doing this, the emulsifiers took the form of two easily-available (produced in Poland) non-ionic surfactants, namely Rokafenol N8 and Rokacet S7. An analysis of physico-chemical parameters was carried out and the heat of combustion of the oils and fuel mixes was determined.

The mixing of wet sediments and fuel oil (where appropriate also with waste oil) was done in laboratory conditions and tests carried out on burning in a tubular furnace in an air atmosphere.

Analysis of the waste gases produced during the experiment was performed with the aid of an analyser for oxides of nitrogen, a gas chromatograph and traditional methods (e.g. SO_2 by the iodometric method and NO_2 by spectrophotometry).

Also run at the Rejowiec Cement Works were technical tests on the combustion of the sewage sludge (via supply to a cement kiln in clinker slurry) and of its mixtures with fuel oil. Monitoring of this, and of its influence on the technological process, was done via analysis of flue gases, particulates from the electrofilter and clinker. ASA and ICP did the analysis of solid samples after prior mineralisation of the material in a microwave mineraliser.

3. CHEMICAL AND ENERGETIC CHARACTERISTICS OF THE SEWAGE SLUDGE

There can be no doubt that the attractiveness of sludge as a fuel increases with its calorific value. Equally, what is important from the environmental point of view is the chemical composition of the sludge, and most especially the content of those components that could be found in the waste gases from the cement kiln. Table 1 details the average chemical composition of sewage sludge.

The dry-mass content of organic substances varies within the range 52.6-58.4%, thereby ensuring a positive energetic balance for the sludge.

Table 2 sets out the results of work on the heat of combustion of wet sewage sludge and sludge dried at a temperature of 105°C. The heat of combustion of the dried sediment varies across the narrow range 14.43-15.52 MJ/kg. The dried sludge is thus a fuel comparable in energy terms with wood, whose calorific value varies in the range 14-19 MJ/kg. The stability of the heat of combustion of the dried sludge should be stressed, an element that is undoubtedly connected with the practically constant level of organic substances.

Table 1. Chemical composition of sewage sludge

Parameter	Unit	Content	Range (10 samples)
CaO	g/kg dry mass	384	299 – 412
Fe_2O_3	g/kg dry mass	188	142 – 208
SiO_2	g/kg dry mass	139	102 – 156
S	g/kg dry mass	19.4	15.2 – 24.6
P_SO_5	g/kg dry mass	93	46 – 157
K_SO	g/kg dry mass	18.3	14.6 – 21.8
Al_2O_3	g/kg dry mass	22.4	16.8 – 28.4
H_SO	%	77	75.0 – 79.2
Ash (600°C)	%	9.74	8.63 – 11.4

Table 2. Combustion heats of wet and dried sewage sludge

	Unit	Heat of combustion	Range (9 samples)
Wet sludge	MJ/kg	1.27	0.89 – 1.61
Dry sludge	MJ/kg	14.89	14.43 – 15.52

The energy required to dry 1 kg of wet sewage sludge is proportional to the amount of water, and varies between 2.02 and 2.12 MJ/kg of wet sediment in the case of sludge from Hajdów.

The total calorific value of the dry matter contained in 1 kg of wet sludge is greater than the amount of energy needed to evaporate off the water. This therefore assures the small positive energetic balance that is capable of being obtained in the process of burning sludge. The calorific value of the wet sludge varies in the range 0.89-1.61 MJ/kg.

It would be hard to regard mechanically-dewatered sewage sludge as an attractive fuel, but the linkage of its utilisation with use in the firing of a cement kiln allows the process to gain justification on economic grounds.

4. THE GENERATION OF MIXES OF WET SEDIMENTS WITH FUEL OIL AND WASTE

A mix of sludge and oils must be persistent enough to ensure undisrupted transport over the distance of c. 50-km between Lublin and Rejowiec, and subsequent feeding into the rotary kiln of a uniform fuel mix.

The parameters, which would seem to determine the suitability of the mix as a fuel, are:

- the calorific value;
- permanence - measured as the time that passes before the aqueous and organic layers separate out;
- viscosity - influencing the possibilities for supplying the mixture via a lance or burner pipe.

The work was carried out on mixtures with 27.5, 37.5 and 47.5% of wet sludge and up to 5% of each of the surfactants (separately or simultaneously). The generation of the mixes was achieved in a laboratory homogeniser with the number of turns of the cutting blades set to 500 revolutions per minute and a mixing time of 10 minutes.

4.1 Mixes of Sewage Sludge and Fuel Oil

The trials made use of the widely available EKOTERM fuel oil from the Płock Petrochemical Works (*Zakłady Petrochemia Płock S.A.*). This has a calorific value of 43.13 MJ/kg, while the chemical composition is as presented in Table 3.

Table 3. Physico-chemical parameters of fuel oil

Parameter	Unit	Result
Density in 20°C	kg/m^3	0.85
Dry mass	%	< 0.01
Water	%	< 0.03
Ash (600°C)	%	0.027
S	%	1.57
Cl	%	< 0.01
Combustion heat	MJ/kg	43.13

The minimal calorific value for a fuel mix generated from sewage sludge that could assure the undisturbed operation of the cement kiln was 16 MJ/kg. This points to a lower limit for the content of fuel oil of c. 40%. The results of the studies carried out are presented in Table 4.

Table 4. Parameters of fuel obtained by mixing sewage sludge with fuel oil.

Content of sewage sludge	Content of surfactant [%]		Stability	Viscosity	Combustion heat
[%]	Rekafenol N8	Rokacet S6	[day]	[°E]	[MJ/kg]
27.5	-	5	hours	203	30.49
27.5	5	-	hours	183	30.56
27.5	2.5	2.5	4	163	30.58
37.5	2.5	2.5	5	179	25.26
47.5	2.5	2.5	4	192	21.51
47	1	5	2	169	21.63
47	5	1	1	161	21.43

The most favourable parameters are those of the mix of sewage sludge and fuel oil containing 2.5% each of the two surfactants. These mixtures persist for 4-5 days, making their longer-term preservation possible. The mix containing 47.0-48.5% of sludge was characterised for research on a technical scale. While it did have considerable viscosity, this did not prevent it from being introduced into the rotary kiln with the aid of a burner or lance.

4.2 Mixes of Sewage Sludge and Waste Rolling-mill Oil

The waste oil was generated at the Sendzimir Steelworks in Nowa Huta, in quantities of some 50-120 Mg/month. The calorific value of the oil was c. 35 MJ/kg, making it a product worse than fuel oil in energetic terms. It contains a number of pollutants (Table 5) in the form of sulphur (0.5-4.0%) and heavy metals like Pb, Cd, Cr, Zr, Mo, Ni and Cu.

Table 5. Physico-chemical parameters of waste oil generated by the rolling mills

Parameter	Unit	Result
Density	kg/m^3	930 – 1080
Viscosity (80°C)	°E	6.5 – 7.9
Water	%	18.2 – 21.3
S	%	0.5 – 4.1
Solid substances	%	5.3 – 8.1
Ash	%	6.8 – 7.3
Cu	mg/kg	420 – 640
Cd	mg/kg	< 5
Cr	mg/kg	89 – 112
Pb	mg/kg	23 – 48
Ni	mg/kg	110 – 150
Zr	mg/kg	205 – 355
Mo	mg/kg	19 – 25
Combustion heat	MJ/kg	34.8 – 35.6

The calorific value of the waste rolling-mill oil denotes a minimum value for the content of waste oil of c. 50%. The results of the research carried out are presented in Table 6.

Table 6. Parameters of fuel obtained by mixing sewage sludge with fuel oil.

Content of sewage sludge	Content of surfactant [%]		Stability	Viscosity	Combustion heat
[%]	Rekafenol N8	Rokacet S6	[day]	[°E]	[MJ/kg]
27.5	-	5	hours	234	25.23
27.5	5	-	hours	225	25.64
27.5	2.5	2.5	4	153	25.71
37.5	2.5	2.5	5	163	22.73
47.5	2.5	2.5	4	189	18.62
47	1	5	2	194	19.28
47	5	1	1	174	18.19

As in the case of mixtures with fuel oil, the most favourable parameters are those of the mix of sewage sludge with waste oil and 2.5% of each of the two surfactants.

5. TECHNICAL TESTS ON THE UTILISATION OF THE SEWAGE SLUDGE AT CEMENT PLANT

On May 25, 1999, c. 21 Mg of sewage sludge was supplied to the Rejowiec Cement Works. The next day, technical tests on the utilisation of the sludge were carried out, by (1) adding sewage sludge to clinker (Portland) slurry in the amount of 2%, and (2) burning the mixture of sludge with fuel oil.

5.1 Combustion of the Mix of Sewage Sludge with Fuel Oil

The test burning of the fuel mix was carried out in furnace no. 1. This furnace has a mounted lance to allow for the introduction of liquid wastes. While it is true that the emulsion obtained in the laboratory showed a high degree of viscosity, no problems were noted with the choice of parameters for its introduction into the cement kiln (Table 7). The lance is suspended beneath the burner for coal dust in such a way that the fuel or liquid wastes are introduced into that zone of the furnace with a temperature of c. 1500°C.

Table 7. Technical parameters of experiment

Pressure of fuel mixture at lance outlet	0.5 MPa
Pressure of compressed air at lance outlet	0.5 MPa
Velocity of mixture serving to cement kiln	1.1 m^3/h
Time of experiment	8 h

Table 8. Physico-chemical parameters of fuel oil and its mixture with sewage sludge

Parameter	Unit	Fuel oil	Mixture with sludge
Density (20°C)	Kg/m^3	0.981	0.996
Dry mass	%	< 0.01	14.73
Water	%	< 0.03	23.62
Ash (600°C)	%	0.027	10.35
S	%	1.57	1.03
Cl	%	< 0.01	0.06
Combustion heat	MJ/kg	43.13	25.33

The fuel mix was produced in the waste tank by mixing 5 Mg of sludge with 5 Mg of fuel oil and 200 kg each of the non-ionic surfactants Rokafenol N-8 and Rokacet S-7.

The samples of fuel obtained were analysed, as was the clean oil, and the parameters presented in Table 8 were determined.

On 26.05.1999, all the furnaces used engaged in the production of the Portland clinker whose chemical composition is set out in Table 9.

Table 9. Chemical composition of portland cement and coal dust in the day of experiment.

Parameter	Unit	Result
SiO_2	%	13.65
Al_2O_3	%	2.50
Fe_2O_3	%	1.24
CaO	%	45.15
Ash in coal dust	%	13.25
Combustion heat	MJ/kg	24.94

The flammable component of the fuel comprised oil-derivative hydrocarbons - elements of the fuel oil as well as organic matter from the sewage sludge. The main products of their combustion are CO_2 and H_2O. It is obvious that the substitution of a solid fuel by a liquid or emulsion will reduce the emission of CO_2 to the atmosphere, as well as limit the conditions under which oxides of nitrogen may be generated.

The mineral residue from the combustion of the fuel mixture, and especially from the sewage sludge, amounted to 10.35% in the experimental conditions. In the coal dust the ash content may reach 18%. In technological terms, these differences are not of practical significance, since in both cases the ash is absorbed by the raw-material set and becomes part of the composition of the clinker.

5.2 The Utilisation of Sewage Sludge Through its Addition to Clinker Slurry

The addition of sewage sludge to clinker slurry was carried out in furnace no. 3. The sewage sludge was added at the rate of 2% gives rise to only limited changes in the latter's chemical composition, in particular increasing the amount of volatile substances somewhat, while the mineral components of the slurry are diluted to some extent (as presented in Table 10).

The process of utilisation was run for 8 hours. In the course of the experiment, waste gases, dust and cement were analysed.

Table 10. Chemical composition of portland sludge without and with sewage sludge.

Parameter	Unit	No. Sewage sludge	+2% of sewage sludge
SiO_2	%	13.65	13.41
Al_2O_3	%	2.50	2.47
Fe_2O_3	%	1.24	1.23
CaO	%	45.15	44.31

5.3 The Influence of Sewage Sludge on the Clinker Burning Process, the Emission of Pollutants and the Quality of Cement

Measurements of particulate emissions and of the composition of discharged gases were carried out on 24, 26 and 27 May 1999, i.e. two days before the experiment, on the day of the experiment and on the day after it had been carried out.

There was no sign of virtually any negative processes, which would have led to increased emissions of gases and particulates. However, it was possible to note a reduction in emissions of particulates and sulphur dioxide from furnace no. 3 fed with clinker slurry containing 2% sewage sludge, while the level of emission of oxides of nitrogen was at an unchanged level.

On the basis of the results obtained it is not possible to state categorically whether the observed reduction in emissions of particulates and SO_2 is linked directly with the supply of sewage sludge to the furnace. Further experiments will need to be carried out with a view to accounting for the phenomenon.

The cement kiln fed with a mix of fuel oil and sewage sludge was not shown to produce greater emissions of particulates, oxides of nitrogen or sulphur dioxide.

The emissions of particulates and gases referred to be detailed in Tables 11-13.

Rotary kiln no. 1 showed almost no fluctuation in particulate emissions under the influence of the dosing of the mix of sewage sludge and fuel oil. Similarly, emissions of oxides of nitrogen did not change. On the other hand, the amount of sulphur dioxide emitted to the atmosphere declined slightly (by c. 15%).

Kiln no. 3 reported a clear, 1.5-2-fold, reduction in particulate emissions and a c. 20% fall in those of sulphur dioxide. In contrast, the amount of nitrogen dioxide in no way depended on whether or not sewage sludge was being fed into the rotary cement kiln.

Table 11. Emission of dust at the time of experiment

Day	Emission	Intensity of gases	Temperature of gases
	kg/h	m^3/s	K
Cement kiln No 1			
2 days before	8.46	33.6	448
0	8.32	32.1	443
1 day after	8.13	33.7	446
Cement kiln No 3			
2 days before	9.24	35.1	443
0	5.29	34.7	443
1 day after	7.69	34.9	444

Table 12. Emission of SO_2 at the time of experiment.

Day	Emission	Intensity of gases	Temperature of gases
	kg/h	m^3/s	K
Cement kiln No 1			
2 days before	8.32	33.6	448
0	6.78	31.1	443
1 day after	8.14	33.7	446
Cement kiln No 3			
2 days before	10.57	34.9	443
0	7.88	34.7	443
1 day after	9.46	34.9	444

Table 13. Emission of NO_x at the time of experiment

Day	Emission	Intensity of gases	Temperature of gases
	kg/h	m^3/s	K
Cement kiln No 1			
2 days before	17.61	33.6	448
0	16.46	32.1	443
1 day after	16.28	33.7	446
Cement kiln No 3			
2 days before	18.64	34.9	443
0	19.02	34.7	443
1 day after	19.26	34.9	444

One experiment cannot be taken as providing a basis for conclusions to be drawn regarding whether the observed decline in emissions of particulates and sulphur dioxide is due to the modified burning conditions in the cement kiln resulting from the addition of sewage sludge and fuel oil. On the other hand, an undoubtedly unfavourable effect was found to arise both during the combustion of the fuel-oil/sewage sludge mix and during the addition of sludge in clinker slurry. What is involved here is the emission of odoriferous substances detectable organoleptically. The reduction in emissions of these substances to the atmosphere will require the modernisation of the cycling of waste gases from the rotary kilns.

The chemical and phase composition of the clinker from furnaces 1 and 3 is presented in Table 14.

Table 14. Chemical and phase composition of portland clinker

Compound		2 days before	0	1 day after
Cement kiln No 1				
SiO_2	%	22.42	22.52	22.18
Al_2O_3	%	5.24	5.13	5.28
Fe_2O_3	%	1.99	2.08	2.07
CaO	%	66.98	67.70	67.8
MgO	%	0.72	0.69	0.82
SO_3	%	0.31	0.45	0.38
CaO free	%	1.03	1.15	1.18
C_3S	%	60.05	59.90	60.24
C_2S	%	19.42	19.37	19.51
C_3A	%	9.96	10.07	9.66
C_4AF	%	6.28	6.32	6.42
Cement kiln No 3				
SiO_2	%	22.06	21.66	21.98
Al_2O_3	%	5.04	5.23	5.16
Fe_2O_3	%	2.09	2.08	2.03
CaO	%	65.83	66.43	67.56
MgO	%	0.68	0.71	0.78
SO_3	%	0.38	0.42	0.43
CaO free	%	1.21	1.13	1.06
C_3S	%	61.02	60.15	60.69
C_2S	%	19.87	19.63	19.42
C_3A	%	9.24	9.56	9.98
C_4AF	%	6.14	6.21	6.18

The observed variation in the chemical and phase composition of the Portland cements are limited, while it needs to be stressed that the utilisation of the sewage sludge had no influence whatever on the quality of the product obtained. This gains further confirmation in the studies of properties of the cements.

5.4 The Influence of Sewage Sludge Combustion on Fly Ash

Important from the point of view of the regulations concerning the protection of the environment against wastes is the determination of the amounts of heavy metals contained in ash. Table 15 presents comparative data on the contents of heavy metals in ash from the burnt coal and the mixture of sewage sludge with fuel oil. These allow the fuel mix to be considered to introduce a lesser amount of heavy metals than the coal dust.

Table 15. Heavy metals content in coal ash and ahs from mixture of fuel oil with sewage sludge.

Metal	Coal ash		Fuel oil + sewage sludge
	2 days before	day of experiment	
	[mg/kg]		[mg/kg]
Zn	181	136	164
Cu	68.3	110	25.8
Cd	18.6	36.4	8.9
Cr	63.9	42.3	17.3
Pb	112	96.5	9.6
Ni	69.1	91.8	10.4

The above compilation is thus unambiguous in indicating that the utilisation of sewage sludge in cement kilns has no negative effect on the management of wastes by a cement works.

6. SUMMARY

The combustion of sewage sludge in rotary cement kilns and their provision as an additive to clinker slurry does not bring about the increased emission to the atmosphere of particulates, sulphur dioxide or nitrogen oxides, and does not have a negative influence on waste management in a cement works.

The generation of a persistent mix of sewage sludge and either fuel oil or waste rolling-mill oil of appropriate calorific value does not constitute a more major problem. From the economic point of view, it would be worth recommending the production of mixes at the place of combustion.

Excessive (perceptible) amounts of odoriferous substances arise as sewage sludge are utilised. To prevent the emission of odours to the atmosphere it will be necessary to modernise the cycle for waste gases discharged from the rotary kilns.

The costs of the energy obtained from the mix of sewage sludge with fuel oil exceed at least twofold the cost of the energy obtained from coal dust. At the present time this does not allow the implementation of this solution to be recommended, though this would reflect economic, rather than other, circumstances.

From the point of view of protecting the environment against wastes, the utilisation of sewage sludge and which to utilise and dispose of wastes of this kind should consider waste rolling-mill oils as one of the alternative methods.

REFERENCES

1. Zielewicz-Madej, E., and Fukas-Płonka, Ł., 1995, Kryterium wyboru sposobu przeróbki osadów ściekowych. In: Materiały Konferencji nt. *Współczesne problemy gospodarki wodno-ściekowe*: 513-522. Koszalin-Kołobrzeg.
2. Stępniak, S., 1990, Granulowanie i mineralizacja osadów oczyszczalni komunalnych metodą Hydro-Kalz-2000. *Gaz, Woda i Technika Sanitarna No. 11*.
3. Bzowski, Z., and Zawiślak, J., 1997, Zastosowanie osadów ściekowych na składowiskach odpadów komunalnych. *Ekoinżynierii No.1*: 25-29.
4. Kozłowska, B., 1998, Dlaczego piroliza. In: Materiały Konferencji nt. *Termiczna utylizacja odpadów*. Poznań.
5. Benoit, M.R., 1996, Wykorzystanie odpadów jako paliwa w piecach cementowych. In: Materiały Konferencji nt. *Zintegrowane systemy ochrony środowiska. Termiczna utylizacja odpadów*. SIiTPChem, Szczecin.
6. Bień, J.B., and Bień, J.D., 1998, Unieszkodliwianie osadów ściekowych za pomocą metod termicznych. In: Materiały Konferencji Naukowo-Technicznej *Osady ściekowe w praktyce*. Częstochowa-Ustroń.
7. Nowak, E., 1998, Fizykochemiczne przesłanki obecności i limitowania zawartości pierwiastków akcesorycznych i metali w klinkierze cementowym. Internal Reports of IMMB. Kraków.
8. Pawłowski, L., Kozak, Z., Gierżatowicz, R., and Dudzińska, M.R., 1998, Neutralisation of Hazardous Wastes Combined with Clinker Manufacturing. In: Environmental Science Research vol. 55, Chemistry for the Protection of the Environment 3. Plenum: 165-172.
9. Walaszek, J., and Radomski, A., 1997, Utylizacja odpadów w piecach obrotowych do wypału klinkieru. In: Materiały III Międzynarodowej Konferencji nt. *Spalanie odpadów - technologie i problemy*: 313-325. Szczyrk.
10. Piecuch, T., 1998, Termiczna utylizacja odpadów i ochrona przed szkodliwymi składnikami spalin. Politechnika Koszalińska, Koszalin.

THE ENVIRONMENTAL INFLUENCE OF HEAVY METALS INTRODUCED TO THE CLINKERING PROCESS WITH WASTES

Ryszard Gierżatowicz
Institute of Environmental Protection Engineering, Lublin University of Technology, 40, Nadbystrzycka Street, 20-618 Lublin, Poland

Key words: clinker, heavy metals, waste incineration, metal immobilisation, retention coefficient

Abstract: Utilisation of mineral wastes containg heavy metals in cement kiln is well-know method. Metal immobilisation into clinker might be determined by retention coefficients (Wr). Calculation of Wr enables the estimation of heavy metal emission to the atmosphere. Paper presents the calculation of Wr values for selected metals, based on measurements for full-scale experiments for 'dry' and 'wet' methods.

1. INTRODUCTION

The thermal characteristics of the rotary cement kiln ensure ideal conditions for the safe combustion of a range of wastes in association with the generation of energy. In addition, there is an opportunity to utilise mineral wastes by way of their incorporation into the structure of clinker. Also speaking in favour of the use of the rotary cement kiln in the utilisation of wastes is the fact that its adaptation to this end does not require more major financial outlays.

In the case of the utilisation of wastes containing metals, their secondary emission to the environment requires explanation. To facilitate discussion, the metals in question may be separated into four different classes on the basis of volatility (Table 1).

In the conditions under which clinker burns, the volatile elements enter waste gases much more readily, and then pass with them into the atmosphere. While, the immobilisation of metals in clinker has a number of advantages, oversupply may lead to:

- the production of a portion of the clinker and cement that is worthless from the point of technical and sanitary standards;
- the emission to the atmosphere of amounts of metals in excess of permitted standards.

The metals in question also occur in the natural raw materials used in the production of clinker.

Table 1. Classification of metals bussed on volatility (by Nowak[1]).

Class	
Hig rolatile	Hg
Volatile	Se, Bi, Pb, Tl, Cd
Semi-volatile	Rb, Cs
Low volatile	Se, Nb, Ca, Ba, B, Sn, Ta, As, Cr, Co, U, Cu, Zn, Sb, Mo, Ni, V

2. THE PHYSICO-CHEMICAL BASES TO THE IMMOBILISATION OF METALS IN CLINKER

Clinker is a composite of four basic mineral phases with a small share of formless glassy substance containing trace quantities of CaO, SiO_2, Al_2O_3 and Fe_2O_3, as well as most of the trace elements.

The minerals mentioned in Table 2 are complex configurations of the solid solution type, often of different crystalographic forms. A property of such solid solutions is the possibility for atoms of alien elements - otherwise known as accessories - to be built into their structure. They replace atoms of the basic elements (Ca, Si, Al or Fe) on account of concordances of atomic radii.

Table 2. Clinker components

Mineral	Chemical formula	Technical aberration	Content in Polish clinkers
Alite	$3\ CaO \cdot SiO_2$	C_3S	55-65
Belite	$2\ CaO \cdot SiO_2$	C_2S	15-25
Tricalcium aluminate	$3\ CaO \cdot Al_2O_3$	C_3A	8-12
Brown milleryte	$3\ CaO \cdot Al_2O_3 \cdot Fe_2O_3$	C_4AF	8-12

The accessory elements to be encountered in the mineral phases of clinker are presented in Table 3[1].

Table 3. Accessory elements found in clinker

No	Mineral phase	Element
1.	C_3S	Mg, Al., Na, K, Fe, Mn, Ti, Cr, S, P, Ba
2.	C_2S	Al., Fe, Mg, Na, K, Cr, Mn, Ba, P, S, Ti, As, B, Zr
3.	C_3A	Fe, Mg, Si, Ti, Na, K
4.	C_4AF	Mg, Si, Ti, Mn, Cr

The content of different elements in alite and belite may be looked upon as a particular kind of solubility. The limit, i.e. maximal, solubility of certain elements (as oxides) is presented in Table 4[1].

Table 4. Solubility of elements (as oxides) in alite and belite [w/w %]

Metal oxide	Alite solubility	Belite solubility
CrO_3	1.4	0.8
Ga_2O_3	1.9	nd
MgO	2.0	0.54
Mn_2O_3	0.074	nd
MnO_2	nd	0.06
Al_2O_3	0.7-1.7	2.6
TiO_2	0.6	0.46
Fe_2O_3	0.4-1.6	1.3
BaO	2.25	0.3
Na_2O	0.3	0.11
ZnO	0.6	nd
P_2O_5	2.0	0.17
V_2O_5	0.6	nd
V_2O_3	nd*	0.25
As_2O_3	0.6	nd
K_2O	nd	0.78
SO_3	nd	0.62
B_2O_3	nd	0.3
SrO	nd	0.3
ZrO_2	nd	0.3

* nd - no data available

3. CRITERIA DESCRIBING THE PERMISSIBLE AMOUNTS OF AN ACCESSORY ELEMENT IN A FED RAW MATERIAL

The determination of the permissible amount of an accessory element introduced additionally to fed raw material is conditioned by the following criteria:

- its content in clinker, and in consequence in cement as a final product, may not bring about a worsening of the latter's technical or sanitary properties;
- the level of emission to the atmosphere along with process gases may not exceed the amount permitted by law.

The starting point for the assessment of emissions of heavy metals to the atmosphere is the so-called coefficient of metal retention, which characterises the degree of immobilisation in clinker.

The materials balance for the rotary kiln reveals that heavy metals enter it:

a) with fuel,

b) with raw materials in the form of the slurry of the wet method or the so-called "flour" of the dry method;

c) with the wastes utilised.

These metals leave the kiln in clinker (in which they constitute a structural component) or in the particulates or aerosols emitted to the atmosphere. Small amounts are also incorporated into the ceramic fireproof cladding of the kiln.

The materials balance for a rotary cement kiln may be characterised by the equation:

$$L_p + L_s + L_o = L_k + L_d$$

where:

L_p is the load of metal introduced to the kiln with fuel [kg/h]

L_s the load of metal introduced with raw materials [kg/h]

L_o the load of metal introduced with wastes [kg/h]

L_k the load of metal leaving the kiln in clinker [kg/h]

and L_d the load of metal leaving for the atmosphere (basically as particulates) [kg/h].

The retention coefficient R_f is described by the equation [%]:

$$W_r = [1 - L_{de}/(L_k + L_d)]\cdot[1 - L_{de}/L_p + L_s + L_o)] \cdot 100\% =$$
$$= [1 - L_{de}/L_p + L_{(s+o)})] \cdot 100$$

where:

L_{de} – load of

$L_{(s+o)}$ – load of metal introduced with wastes and raw material

The coefficient characterises the amount of the element that is incorporated into the structure of the clinker. Indirectly, it shows the proportion of the mass of this element in the production wastes that are the particulates emitted to the atmosphere (on the assumption that particulates retained at the dedusting installations are returned to the production process). The retention coefficient is different in the cases of different heavy metals and is dependent on the type of raw materials used in production, the technologies of production and the technical parameters of the waste gases discharged. A knowledge of the retention coefficient allows the size of the hourly input of wastes to be planned, which is to say that - on the basis of a familiarity with the chemical composition of the natural raw materials and wastes - a preliminary determination of the ratio between the mass of wastes and that of natural raw materials in the mix fed into the rotatory kiln can be obtained.

The sizes of the retention coefficients for most heavy metals attest to the very high degree to which they are retained in clinker[2,3,6].

The retention coefficients of some heavy metals obtained in the course of our own studies[4] are presented in Figure 1.

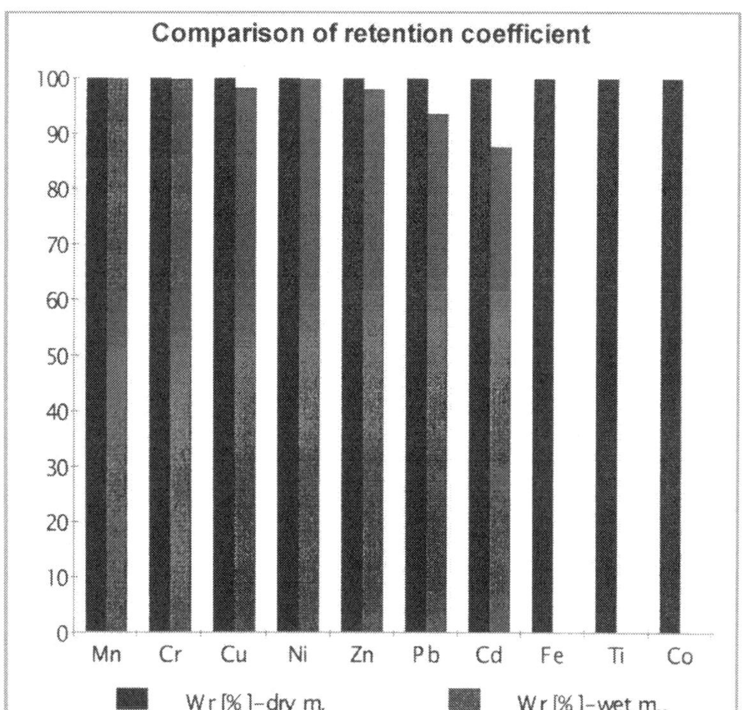

Figure 1. Comparison of Wr values for dry and wet method.

Industrial wastes were introduced into the "wet" process together with natural raw materials via a crusher. Thanks to this they attained a high degree of homogenisation in the furnace sludge. For the "dry" process, the wastes were introduced in the appropriate proportion to stored piles of natural raw materials, after which they were ground up together and dosed following a period of storage in silos via cyclonic heat exchangers to the rotatory kiln. Concentrations of metals were determined in coal, in kiln slurry or raw "flour", in clinker and in the particulates retained by the electrofilters or emitted.

The composition of the wastes introduced into the cement kilns is given in Table 5.

Table 5. Chemical characteristic of wastes [g/kg of wet wastes]

	Waste 1	Waste 2	Waste 3
Water	392.85	711.08	541.4
Ca	54.475	61.57	30.3944
Mg	2.655	nd	3.2114
Na	16.056	nd	4.2584
K	3.991	7	3.3261
Al.	13.223	4	nd
Si	91.821	nd	nd
Fe	47.582	3.26	2.2078
Mn	5.621	87.36	nd
Cr	3.49	nd	1.577
Cu	1.32	0.3	0.0498
Ni	10.315	0.149	0.0728
Zn	7.182	0.1	1.8924
Cd	0.001	4.01	0.0041
Pb	0.037	0.198	5.5627
Ti	11.1	0.032	nd
P	10.478	nd	nd
S	1.444	nd	nd
Cl	1.374	nd	nd
Co	0.265	nd	nd
Sb	traces	nd	nd

The proportion of wastes in the slurry relative to that in the raw-material "flour" was a quantity varying between 2.4 and 6.7%, corresponding to a rate of throughput of between 1.5 and 5 Mg/h.

The obtained values for retention coefficients do not depart markedly from those given in the literature, except for the fact that the retention of such metals as Pb and Cd is much lower (by c. 6.3 and 12.3% respectively) in the case of the "wet" method. Thus these metals (out of the group studied) would seem of crucial importance in assessing opportunities for the "consumption" of metal-bearing wastes and should be monitored by the cement works involved. Their readier volatility in relation to the others may probably be explained by reference to the considerable higher-temperature volatility of the metals themselves, their oxides and their halogenic compounds. However, mention has also been made of their possible emission in the form of aquous aerosols of the appropriate salts (in the wet method).

The emissions of some metals are undoubtedly influenced by the presence of chlorine compounds in the kiln. It is mainly Pb and Cd that are sensitive to the presence of chlorine, while such metals as Zn, As, Cr and Ni would seem not to be sensitive.

The retention coefficients found as above for a given rotary kiln allow for the determination of hourly inputs of wastes that avoid non-compliance with the permissible values for emissions of these metals to the atmosphere. In accordance with Polish regulations on the protection of the atmosphere, the value for emission is dependent *inter alia* on permissible 30-minute concentrations of metals in the atmosphere, i.e. on concentrations in the air.

Having adopted values for limit emissions of metals - i.e. those avoiding non-compliance with permitted concentrations in the air - in the case of a cement works operating on the basis of the "wet" method, it is then possible to calculate limit emission values as given in Table 6.

Table 6. Allowable emission limits for the "wet" method

Metal	Permitted emission [kg/h]
Cu	8.64
Ni	0.054
Pb	1.44
Cd	0.144
Cr	3.24
Zn	16.2

It follows from the aforementioned considerations that the limiting element is lead. For this reason further calculations will refer to this element. On the basis of its retention coefficient (Wr = 93.56%), it is possible to calculate the mass of lead introduced into the kiln which should not lead to a situation in which the permitted emission values are exceeded. This value if of 22.3 kg/h, but from this it is necessary to deduct the amount of lead brought into the process with coal and the natural mineral raw materials. The hourly production of clinker in the rotary kiln under consideration is 74.6 Mg/h, which corresponds to a consumption of 108.06 Mg of mineral raw material in the form of slurry, and 22.38 Mg/h of coal. Accepting that the content of Pb in coal is 0.1 kg/Mg on average, while that in the natural raw materials is 0.09 kg/Mg, the amount of lead introduced into the production process with the raw materials is:

$$22.38 \text{ Mg/h} \cdot 0{,}1 \text{ kg/Mg} + 108.06 \text{ Mg/h} \cdot 0.09 \text{ kg/Mg} =$$
$$= 2.238 \text{ kg/h} + 9.725 \text{ kg/h} = 11.9634 \text{ kg/h}$$

This implies that 10.3 kg/h of lead may be introduced to the clinker burning process with wastes, without there being any non-compliance with permissible emissions. This quantity corresponds to a charge of dewatered sludge "3" from Table 2 equal to 1.85 Mg per hour.

The second restriction on the amounts of heavy metals introduced into the rotary cement kiln is that reflecting the influence on the quality of the cement produced.

Limits on the content of metals in clinker have as their aim the retention of the required technical features of the cement and the non-permitting of any negative impact on components of the environment.

The elements mentioned in Table 4 (in the form of oxides) may - in relation to their proportions in clinker:

- improve the course of the clinkering process in small doses, as well as utilitarian features of the cement;
- be mineralisers of clinker synthesis at limit doses;
- destabilise the structure and composition of clinker, lowering its quality, at above-limit doses.

The metals present in cement like As, Cd, Be and Cr (in the hexavalent form) have an influence on health, though they are only capable of very limited elution from cement. There are references to Cr(IV), Ni or Co giving rise to inflammations or allergic changes in human skin (though this mainly relates to employees having professional contact with cement). A certain view on this issue is provided by the calculations from Benoit[2] on the "acceptable" limits for the contents of certain metals in cement determined for those professionally involved with it, on the assumption that as much as 10% of the metals permeate out of the cement into the surrounding environment of the workplace (Table 7).

Table 7. Allowable limits of metals in clinker

Metal	Permitted level [ppm]
Sb	50000
As	200
Ba	50000
Be	50
Cd	1000
Cr(III)	50000
Cr(VI)	100
Cr total	-
Pb	5000
Hg	5000
Ag	1000
Ta	10000

The accurate determination of limit concentrations of particular elements in cement is a source of major difficulties thanks to the lack of information on synergistic impacts within polymetalic configurations.

The permissible concentrations to be met with in the literature[2,3,5,6] amount respectively to:

MgO - 2%

SO_3 - the less the better, but in general in accordance with quality norms, in principle around 1%. Where the molar ratio $SO_3/(K_2O + Na_20) >1$ emissions of SO_2 in gases occur.

P_2O_5 - 0.5%

Na_2O, K_2O - 1%

TiO_2 - 1% (some studies even suggest as much as 2%)

Zn - 0.5% (0.1% according to Nowak[1])

Cr - 0.5%

SrO, BaO - 900 ppm

F - 2%. Calcium fluoride was once used as a mineraliser at concentrations of up to 1%. Nowak[3] proposes 0.5% in order to exclude the possibility of emissions of fluorine with gases

Total heavy metals - 0.1%

An interesting, coherent and cautious means of establishing concentrations of accessory elements was proposed by Nowak[1]. That author distinguished between the following cases:

a) the most toxic and disease-causing metals, which should not be utilised in a rotary kiln, i.e. Hg, Cd, Tl and Be;

b) the metals As, Co, Ni, Se, Te, Sb, Pb, Cr, Cu, Mn, Pt, Po, Rh, V and Sn, for which permissible concentrations in clinker are determined at the level K3=0.015% by weight, i.e. 150 g/Mg of clinker (when converted for the element).

c) elements non mentioned in point b), whose permissible elemental or oxide concentrations in clinker are denoted by the limit solubility in the different crystallographic phases of clinker. This content, designated by the letter R may be calculated from the formula:

$$R = R_{C3S} \cdot \%(C_3S)/100 + R_{C2S} \cdot \%(C_2S)/100$$

where:

R is the total solubility of the element or elements in clinker in % by weight;

RC3S is the limit solubility in % of metal or the weighted mean of the limit solubilities for the complex of metals which account for more than 0.01% of the waste by weight in the alite phase;

RC2S is as above but in the belite phase;

%[C3S] and %[C2S] are the proportions of the phases in the clinker.

In a case in which several metals are present simultaneously what is needed is the mean weighted (according to the proportions by weight of metal oxides) value from the limit solubilities of these oxides. Limit solubilities are of 0.015% in the case of the accessory metals mentioned in point b), occurring naturally in clinker raw materials, while in the case of metals not occurring naturally in these raw materials it is safer to adopt a value of 0.01%. In doing this, one takes into account the natural content of metals in mineral and energetic raw materials, which is to say the "background".

Thus the masses of wastes containing a given metal or complex of metals which may be introduced to clinker without excessive worry can be calculated using the formulae:

$Q_1 = (0.15 - K_2)/K_1 \cdot 1000$ [kg/Mg]

$Q_2 = (R - R_2)/K_1 \cdot 1000$ [kg/Mg]

where:

Q_1 is the kg amount of wastes containing K_1% by weight of the given metal or complex of metals which may be introduced into 1 Mg of clinker in case b);

Q_2 is the kg amount of wastes containing K_1% by weight of the given element which may be introduced to 1 Mg of clinker in case c).

K_2 is the natural % weight content of the same metal or complex of metals in clinker.

Examples of calculations:

a) monometallic wastes from a tannery containing 8.17% by dry mass Cr_2O_3 = 5.59% by dry mass of Cr.

Other metals in these wastes occur in trace quantities, i.e. at less than 0.01%, and so are not taken into consideration. The natural content in clinker amounts to 0.0112% by weight of Cr_2O_3 = 0.0076% by weight of chromium. The limit proportion of chromium is 0.015% by weight in clinker. The addition of wastes to the basic raw-materials charge should not therefore exceed:

$Q_1 = (0.015 - K_2)/K_1 \cdot 1000 = (0.015 - 0.0076)/5.59 \cdot 1000 = 1.323$ [kg/Mg]

b) wastes containing ZnO at 2.11% by dry weight, Mn_2O_3 at 0.1856% by dry weight and PbO at 0.0145% by dry weight. Together, K_1 = 2.3079% by dry weight.

Conversion to the shares of all oxides gives ZnO = 91.33%, Mn_2O_3 = 8.04% and PbO = 0.63%.

The clinker contains 60% alite and 20% belite. The natural content in clinker of the aforementioned total of oxides is K_2 = 197.1 g/Mg = 0.01971%.

Zinc oxide does not occur among the metals mentioned in sub-point b, for which K3 = 0.015%, and the solubility in the belite phase is not known.

Thus the total (limit) solubility of ZnO is calculated on the assumption that the permissible solubility in the belite phase should not exceed 0.01%:

$R_{ZnO} = R_{C3S} \cdot \%(C_3S)/100 + R_{C2S} \cdot \%(C_2S)/100 =$
$= 0.6 \cdot 600/(100 + 0.01) \cdot 20/100 = 0.362\%$

The oxide of manganese occurs in the natural mineral raw materials, so its limit value is established at a level of 0.015%. The oxide of lead does not basically occur naturally in the raw materials, though for safety's sake a limit value of 0.01% is adopted.

In consequence, the permissible content by weight of the metal oxides taken together amounts to:

$R_{wag} = [(\%ZnO \cdot R\ ZnO) + (\%Mn_2O_3 \cdot 0.015) + \%PbO \cdot 0.01)]/100 =$
$= [(91.33 \cdot 0.362) + (8.04 \cdot 0.015) + (0.63 \cdot 0.01)]/100 =$
$= 33.06 + 1.20 + 0.0063 = 0.343\% = 3430$ g/Mg clinker.

In turn, the mass of wastes introduced should not exceed:

$Q_2 + (R_{wag} - K_2)/K_1 \cdot 1000 = (0.343 - 0.01971)/2.3079 \cdot 1000 =$
$= 140.08$ kg/Mg clinker.

4. SUMMARY

The criterion limiting the content of metals in clinker in relation to technological and sanitary features of cement is a more stringent criterion than that limiting the amount in relation to likely resultant emissions to the atmosphere. This means that adoption of the first criterion is sure to mean no failure to comply with the permissible values for emissions of metals to the atmosphere.

REFERENCES

1. Nowak, E., 1998, Fizykochemiczne przesłanki obecności i limitowania zawartości pierwiastków akcesorycznych i metali w klinkierze cementowym. Internal Reports of IMMB.
2. Benoit, M.R., 1996, Wykorzystanie odpadów jako paliwa w piecach cementowych, in: materiały konferencyjne nt. *Zintegrowane systemy ochrony środowiska. Termiczna utylizacja odpadów*. SIiTPChem, Szczecin.
3. Kurowska, W., 1995, Nowoczesna technologia wytwarzania cementu portlandzkiego a ochrona środowiska naturalnego, in: materiały Ogólnopolskiej Konferencji Naukowo-Technicznej nt. *Przemysł cementowo-wapienniczy a ochrona środowiska*. IMMB No. 16.
4. Pawłowski, L., Kozak Z., Gierżatowicz, R., and Dudzińska M.R., 1998, Neutralisation of Hazardous Wastes Combined with Clinker Manufacturing, in: Environmental Science Research vol. 55, Chemistry for the Protection of the Environment 3. Plenum :165-172.
5. Uchikawa, H., 1993, Aktualne problemy w produkcji cementu, in: *Cement, wapno, gips*. No 3: 81-102.

6. Walaszek, J., and Radomski, A., 1997, *Utylizacja odpadów w piecach obrotowych do wypału klinkieru*, in: materiały III Międzynarodowej Konferencji nt. *Spalanie odpadów – technologie i problemy:* 313-325. Szczyrk.

THE USE OF COAL SPOIL IN THE PROCESS OF CEMENT PRODUCTION

Jerzy Duda
Instiue of Mineral Building Materials in Opole, Poland

Key words: waste fuel, waste utilisation, rotary kiln

Abstract: Significant amount of wastes from different technological processes might be utilising during cement production. High temperature processes of clinker calcination makes the condition preferable for co-incineration of flammable wastes. Paper presents methods and effects of utilisation of coal spoil as substitute for either fuel or raw material in clinker production process.

1. INTRODUCTION

Industrial development and increased consumption have been associated with the generation of large amounts of wastes, of which a limited degree of recycling is possible. Dumping at landfill sites remains the commonest method of proceeding with wastes, but leaving aside the problem of the already-existing shortage of land available for this, there will always be a potential threat to the environment from such sites, irrespective of the safeguards applied. In connection with this, an intensive search is underway for no-waste or low-waste technologies, as well as for methods by which to utilise wastes. An industry capable of using industrial or municipal wastes without any injurious impact on product quality or the environment is the cement industry. The process by which cement is produced is one of a small group that do not generate wastes of their own, while using up considerable amounts of those produced by other industries. Fly ash from power stations is a widely-used additive to cement at cement works, as are blast-furnace slag and REAGIPS from the desulphurisation of waste gases. Combustible wastes form a particular kind whose use the cement industry is interested in, while the generally high degree of interest the industry shows in wastes reflects a desire to lower production costs, and a consequent need to become involved with the difficult problem of waste utilisation. The development of new dry technologies for combustion in kilns using a cyclone heat exchanger and system for preliminary decarbonisation creates opportunties for fuels of lower calorific value or combustible wastes to be burnt. Figure 1 provides a breakdown for the temperatures of gases and material in a kiln employing the dry method, as well as detailing the dosing points for different combustible wastes.

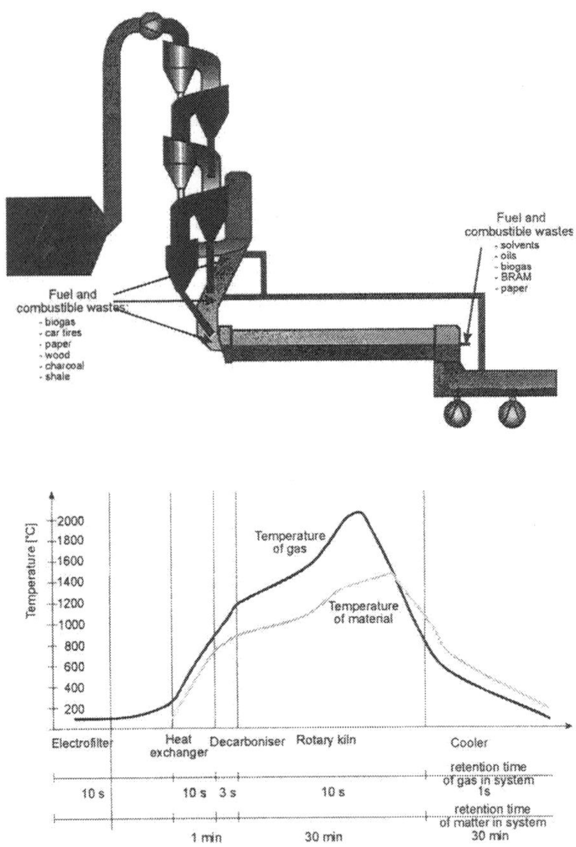

Figure 1. Breakdown of temperatures of gases and matter in a dry-method kiln, and dosing points for different combustible wastes.

Favourable conditions for combustion have ensured the use in the highly-developed countries of used car tyres, spent oil and other combustible wastes as an additional fuel in the decarbonisation process. The high-temperature clinker-burning process and large "combustion chamber" ensure the appropriate time and good mixing of combustion products with the predominating alkaline atmosphere, thereby providing for full compliance of the rotary kilns with EU Directive 92/C130/01, an essential condition if the combustion of hazardous wastes is to be environment-friendly.

2. COAL SPOIL

Mines' exploitation of seams ever poorer in coal is ensuring a steady increase in the amounts of spoil, i.e. coaly shales, being directed to spoil heaps. These wastes constitute a big problem for the mines themselves, as dumping on spoil heaps requires both large outlays and large areas of land, neither of which are at the mines' disposal. The coaly shales accumulating on spoil heaps are 10-15% diffuse coal and therefore pose a threat to the environment linked with the release of the harmful gases SO_2 and CO thanks to spontaneous combustion. In connection with this, mines are actively seeking meethods of placing these wastes with other branches of industry for their further use. On account of its chemical and mineralogical composition and carbon content, coalmining spoil is a very interesting raw material for the cement industry. However, in relation to origin (processing), spoil may differ markedly in terms of the above features. In relation to type and place of generation, coal spoil may be divided into processing and mining wastes, while a further group comprises the so-called burnable shales, i.e. those which undergo combustion in landfills or spoil heaps. On account of their great uniformity of chemical composition, carbon content and physical properties, it is processing wastes that have found the main application in the cement industry. In the wet technologies for which the presence of carbon in the raw material is undesirable, use may instead be made of such burnable shales. In practice, only mining spoil, with its wide variability of chemical and mineralogical composition and contamination by pollutants, may not be used in the production of cement. Otherwise, it is the costs of (long-distance) transport and of the further necessary processing which may rule out the use of coal spoil.

The energy-intensive and large-scale nature of the cement-making process ensures the cement industry's use of considerable amounts of fuel and natural raw materials like limestone, marl and clay. The intensive exploitation of these materials has led to exhaustion of supplies, especially in the case of marls. Many cement works in Poland are now complaining about the lack or low quality of these, while each accessing of new deposits is associated with heavy financial outlays, the taking-over of agricultural land and the degradation of the environment. Another issue attracting little attention to date has been that concerning the rational use of natural raw materials. For example, the high-quality clay whose shortage is bemoaned by the ceramics industry is often used as a raw material in the clinkering process. The arising shortages of traditional raw materials for the production of cement clinker, and the need for rational management of natural raw materials combines with the need for coal spoil to be utilised to encourage the extensive pursuit of research into opportunities for using these wastes as a raw material in the production of cement. The mineralogical and chemical composition of some coal spoil corresponds to those of the so-called base raw materials used in the cement industry. In connection with this, work was carried out as early as in the 1970s to use wastes from hard-coal mines in the production of cement. In the course of the burning of the so-called shale slurry generated from hard coal and coaly shale, no significant differences compared with traditional slurry were noted, while the obtainment of clinker of a 40 MPa physical strength confirmed that such coal spoil might be used in the process of cement-production. However, in spite of these promising results, coal spoil did not find wider application, mainly as a consequence of the technological difficulties connected with obtaining a uniform slurry. These in turn reflected the less favourable slurrying and milling properties of shale as opposed to limestone. Further major problems were posed by the diffuse coal to be found in the wastes since, in burning within the cold part of the kiln, this brought about the drying of the raw material, as well as the breaking-apart of the granules that form. The result of the latter process was a major increase in the dust-loading of waste gases from the furnace, which gave rise to considerable difficulties with dedusting and the later utilisation of particulates from the filters. This all led to a search for another way in which coal spoil might be used to produce clinker. As a result, *inter alia*, of the content of coal in wastes, attempts were made to use them as a raw material and fuel in the clinkering process. Such conditions arose with the development of new dry-method combustion technologies with the system of preliminary decarbonisation.

3. A REVIEW OF METHODS APPLIED IN THE UTILISATION OF COALY SHALES

The way in which coal spoil is used depends mainly on the technologies of clinker production, as well as on the chemical composition and carbon content of the waste material. The energetic effect of the use of wastes containing a combustible component is dependent on the manner in and point at which these wastes are dosed. The full heat gain from the combustion of wastes is only achievable where these are introduced to the first stage beyond the furnace relative to the calciner. The simplest way of employing coal spoil is to prepare the traditional "flour" of raw materials comprising limestone and shale as a base raw material; and possibly also the required additives. This method may be applied to a limited extent only, for wastes not containing a combustible component, in the form of the so-called burnable shales. The use of shales containing diffuse coal in this way may lead not only to technological problems but even to an explosion in the electrofilter. This method has been used for many years by the Japanese firm IHI, with coaly shales or bituminous shales of low carbon content. The shales replace clay in this technology and are comminuted along with the other raw-material components. On account of the danger of a NAPIEK occurring in the cyclone, as well as a rise in the temperature of flue gases, the proportion of coal spoil may not be high. A method more favourable than the mixing of spoil with the raw material is separate dosing into the kiln from the hot end, in parallel with the fuel (Fig. 2)[1].

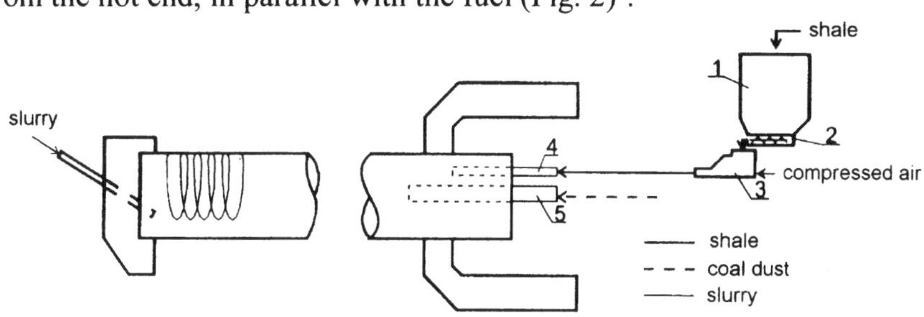

Figure 2. Scheme for the use of coaly shale in wet-method kilns.

This method has a series of limitations. Taking into account the danger that rings of clinker may arise from the shale ash, the amount of coal spoil dosed converted to ash may not exceed 50-75 kg of ash/Mg of clinker. The application of a higher degree of milling (R88 < 10%) and appropriately greater speed may increase the amount of shale dosed to 150 kg of ash/Mg of clinker. A further disadvantage of the method is that - depending on the content of calcium carbonate in the shale - there may be a considerable lowering of the flame temperature in the sintering zone. Only the ROHRBACH method (named after its creator and introduced at the Dotternhausen cement works) will permit the safe and effective utilisation of shale[2]. In this process, use is made of bituminous shales as a substitute raw material and heat source in the process of decarbonisation. This technology entails not only the separate preparation of limestone flour and shale flour, but also the balancing of the mixture of raw materials in the process of combustion within the heat exchanger. Figure 3 presents the kiln line in which the technology has been applied. The limestone flour is dosed as usual between stages 1 and 2 of the exchanger, while the shale flour is introduced to the shaft of the calciner. The heat generated from the combustion of the bituminous shales provides part of that needed in the decarbonisation process, while the ash from the shales constitutes a base raw material which is mixed in the 4th stage with heated limestone flour.

The moisture arising which is deposited in the 5th stage is transferred to the kiln where the final process of clinker burning takes place. The method in question became a prototype solution for later systems of preliminary decarbonisation as well as for many original means by which coil spoil can be used in the clinker-burning process.

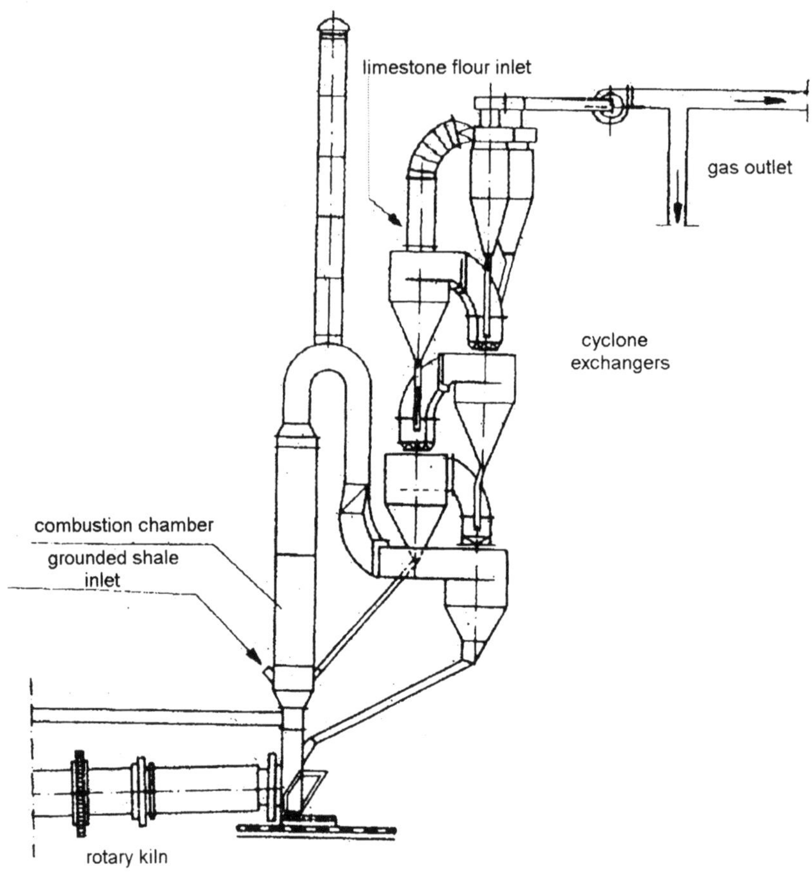

Figure 3. Diagram of the Rohrbach process.

In connection with the planned modernisation of the industry and present shortage or poor quality of base raw materials, research work done at the Institute of Mineral Building Materials (IMMB) has been concerned with the effective use of natural raw materials and the opportunities for use to be made of wastes. The need to employ waste raw materials in the process arose as a concept for the modernisation of the WARTA II cement works was drawn up. It emerged at this time that the works was threatened by a lack of base raw materials, as the previous suppliers at the Łęczyca mine could not assure future supplies on account of the working-out of its deposit and high extraction costs. In connection with this, the works was faced with a need to open its own mine, with all the heavy associated outlays and likely degradation of the natural environment. In such a situation, consideration began to be given to the devising of a technology that would allow coal spoil to be used as a raw material. Figure 4 presents the method for modernising the kilns at the WARTA II cement works in line with the technologies devised at the IMMB[3]. This method entails the separate preparation of limestone flour and flour from shale wastes. The kiln is supplied at two points: limestone flour is dosed into the conduit linking stages 1 and 2 of the exchanger, while the shale flour passes into the decarboniser.

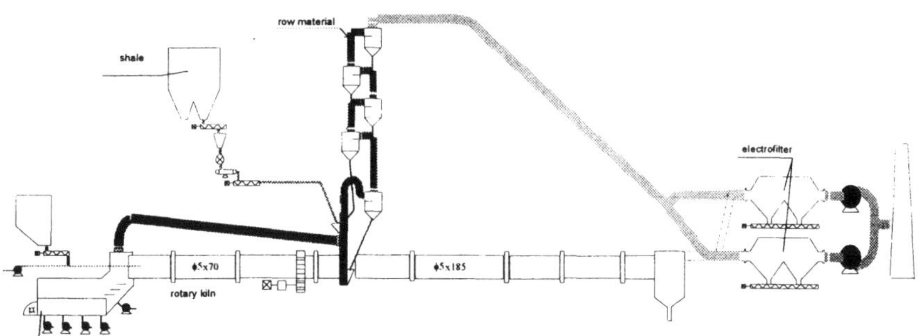

Figure 4. Diagram showing the modernisation of the long kilns from the wet to the dry method.

Table 1 presents the technological parameters of the kiln. In spite of very promising indices for the consumption of heat and low costs of production, the solution has not been implemented on account of the high costs of the odernisation in the face of a lack of funds on the part of the cement works.

Figure 5 presents a flow sheet with technological parameters for a similar solution using coal spoil which was introduced during the period at the RAWANG cement works in Malaysia. The operational data reveal this to be a technology with a heat consumption from the fuel c. 40% lower than in similar dry-method kilns, at 1900 kJ/kg. The solution we propose would have a rather higher heat consumption of 2400 kJ/kg, because the limestone raw material would not allow for a greater proportion of shale in the batch.

Table 1. The technological parameters of the kiln.

No.	Parameter	Measure	Value
			Modernised line
1.	Rotary kiln:		
	- length	m	70.0
	- diameter	m	5.0
	- revolutions	l/min	1.0-2.0
	- motive power	kW	320
2.	Rotary cooling bed:		
	- length	m	47.0
	- diameter	m	5.0
	- revolutions	l/min	1.0-2.5
	- motive power	kW	320
	Grate cooler:		
	- area of bed	m^2	83.5
	- rated output	Mg/h	83.3
	- output after modernisation	Mg/h	104.0
3.	Theoretical index of raw-material consumption, including:		
	- of limestone flour	kg/kg kl.	1.38
	- A – of coal spoil	kg/kg kl.	0.115
	- B – of fly ashes	kg/kg kl.	0.137
4.	Unitary heat consumption:		
	A – total		
	A – from fuel	kg/kg kl.	3,200
	A – from shales	kg/kg kl.	2,420
	B – total fuel	kg/kg kl.	780
		kg/kg kl.	3200
5.	Fuel (coal dust):		
	- calorific value	kJ/kg	23640
	- consumption		
	- A	kg/kg	0.102
	- A	Mg/h	10.7
	- B	kg/kg	0.135
	- B	Mg/h	14.20
6.	Clinker		
	- output	Mg/h	104.2
		Mg/d	2,500
		Mg/year	750,000

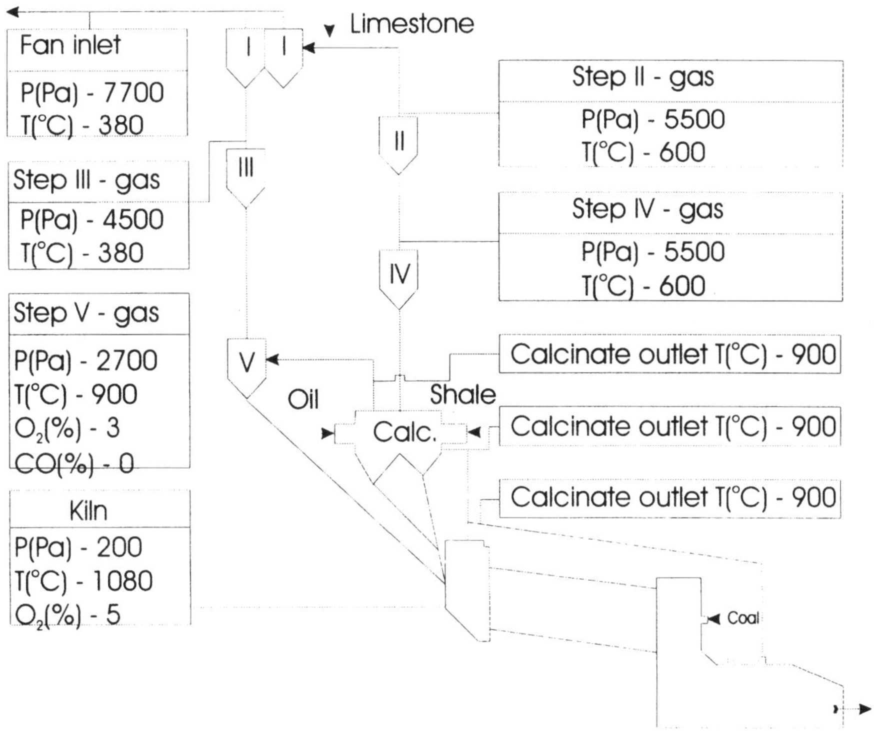

Figure 5. Flow sheet with operating parameters for a kiln at the RAWANG cement works.

4. MEANS OF USING COAL SPOIL WITH THE SEMI-DRY METHOD

The optimal solution for the WARTA II cement works would be the adaptation of the kilns for the dry method using coal spoil. However, in the light of the firm's lack of the required capital, the cost of this modernisation (linked mainly with the building of a new division preparing the dry raw material) makes completion of the task impossible, In connection with this, the IMMB has devised a so-called "intermediate variant" - a semi-dry method not necessitating construction of a new division to prepare dry limestone raw material. The process may later - once funds are assured - be upgraded to the dry method proper. In the semi-dry method an original means of preparing the raw material is applied. The so-called limestone slurry is prepared wet in the existing mills, while the base raw material in the form of coal spoil is prepared in the form of a dry meal in a newly-built shale mill. The whole shale division comprising a pre-homogenisation store, the mill itself and homogenate reservoir has been designed with future use of the dry method in mind. Figure 6 presents the flow sheet for the kiln using the semi-dry method[4].

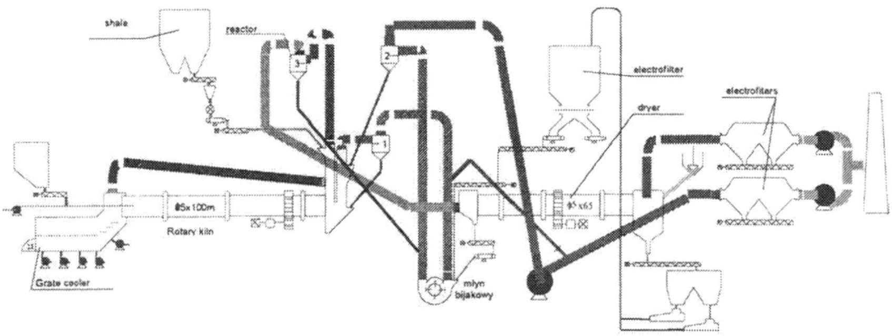

Figure 6. Flow sheet for kiln using the semi-dry method.

The modernisation of the kiln entails the moving of the central part of the drum about 20 m. The lower part with cooler comprises the kiln, while the upper, following the installation of a drive, becomes the slurry drier. The kiln has a 2-stage cyclone heat exchanger and a so-called shale cyclone reactor. After being supplied to the kiln, the limestone slurry is dried preliminarily in the rotary drier by a mix of gases arising from the combustion of shale in the reactor as well as exhaust from the kiln. The final drying and grinding of the granules arising from the slurry in the drier is achieved in the ventilator mill installed between stages 1 and 2 of the cyclone exchanger. The dry meal generated by this process is mixed with ash from the combustion of the shale flour and the mixture obtained is subject to further homogenisation and thermal processing in the cyclone exchanger. Thanks to the fact that the coal-spoil component of the raw material fed to the kiln is in the form of a dry meal, the overall content of water in the material supplied is reduced by c. 10%. An additional gain of energy results from the heat generated in the combustion of shale. The applied semi-dry technology using coal wastes thus has an energetic effect entailing a lowering of the unit heat consumption by c. 2100 kJ/kg of clinker in total, as well as a more favourable impact on the natural environment. Besides resolving the difficult problem of how to utilise coal wastes and hence to eliminate spoil heaps, the technology devised also allows for continued operation of the cement works without any need to open up a new quarry for the base raw material, again with attendant saving of money and a reduced environmental impact (retention of agricultural land).

5. SUMMARY

Previous experience worlwide and of our own has fully confirmed the suitability of using coal wastes and bituminous shales in the cement production process. The energetic gains combine with the high quality of the clinker obtained and the ecological benefits accruing from the retention of agricultural land and limitation of emissions of noxious gases from spoil heaps to provide a sufficient argument in favour of these technologies being developed. Thanks to natural conditions allowing its process to utilise waste raw materials and fuel, the cement industry is able to serve a valuable role in the protection of the environment.

REFERENCES

1. Parisis, J., 1977, VDZ Congress 77:575.
2. Rechmeier, H., 1970, Der fünfstrufige Warmetauscherofen zum Brennen vom Klinker aus Kalkstein und Ōlschiefer. ZKG, No 23.
3. Weryński, B., and Duda, J., 1984, Sposoby zmniejszania energochłonności produkcji cementu, in: *Cement-Wapno-Gips No.1.*
4. Duda, J., 1994, The New Semi-Wet Process Using Coal Shales in Warta II Cement Plant, *World Cement No. 3.*

SOLID WASTE UTILISATION IN CEMENT KILNS

Theodore J. Reese
President and Founder Cadence Environmental Energy, In.c., Michigan City, IN, USA

Key words: NO_x reduction, waste-to-energy, recycling, profit generating

Abstract: Using solid waste as energy in long cement kilns is the right thing to do. It benefits local economies, local industries, the environment, and saves natural resources. Feeding solid waste to long kilns using a Feed Fork is the right way to do it. This article describes the process and the benefits.

1. INTRODUCTION

As the world community enters the new millennium, it does so with optimism and an increased sense of responsibility. We are optimistic because our knowledge and technology offers us the chance to make the world better for everyone. We carry an increased sense of responsibility because as the world's population increases, the decisions we make effect more people than ever before. Good decisions avoid the mistakes of the past.

As the world's industrial economies strive for greater global prominence, it is tempting for them to push forward with a single-minded, production above all focus. While production is the lifeblood of all growing economies, it is only one component of genuine progress. There can be no progress unless the economy, the environment, and the health and well-being of the citizens all benefit simultaneously.

Thermal Solid Waste Utilisation in Regular and Industrial Facilities
Edited by Pawlowski *et al.*, Kluwer Academic/Plenum Publishers, 2000

2. LESSONS FROM THE PAST

In the 1940's, 50's, and 60's, the United States was driven by a production above all focus. New prosperity demanded more products and created more jobs. As new products were produced, industrial wastes, including hazardous chemicals, were being carelessly discarded in the environment causing widespread pollution. While the country was in the midst of unprecedented economic health, widespread pollution put our physical health at serious risk. Citizen outrage pressured the government to act. In 1970, the US Environmental Protection Agency was formed to help solve the pollution problem. US Congress gave it broad regulatory and enforcement power. Now, with huge fines or even criminal prosecution at stake, waste generators and waste management companies were required to properly dispose of waste according to specific regulatory guidelines.

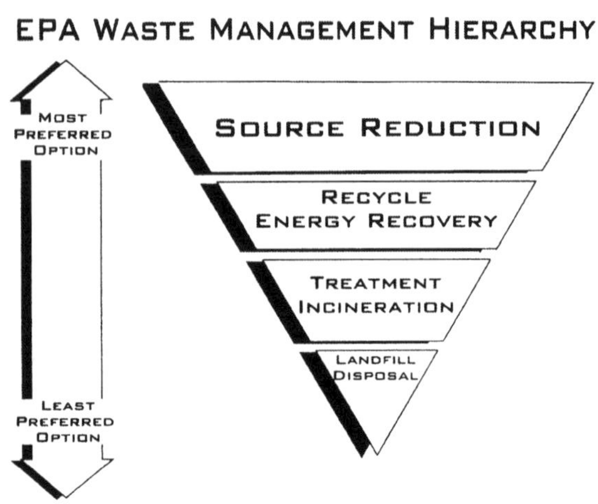

Figure 1. US EPA waste management hierarchy

The US EPA created a four level waste management hierarchy that acted as an overall strategy for dealing with all types of waste. The hierarchy lists waste management options from the most preferable to the least preferable.

This hierarchy became the blueprint that formed the foundation of a nationwide waste management strategy.

As United States industry became more environmentally conscious, one of the most important lessons it learned was this: "*No industrial waste can be simply thrown away without potentially harming human health or the environment*".

Anything carelessly discarded will comeback in another form or place and cause potential harm. Chemical waste discarded on the earth can poison the soil, harm animal and plants and leach into ground water, lakes and rivers and put the health of entire communities at risk. Over the years the wisdom of this hierarchy has spread and successful industrialised countries around the world are adopting the four tier waste management hierarchy as a responsible way to deal with their waste.

3. LOOKING TO THE FUTURE

Experience has shown that the best way to dispose of waste is to use it in a productive and environmentally responsible way. Since nearly 50% of all solid waste generated contains carbon, large scale waste to energy recovery programs have proven to be the most attractive option.

For more than fifty years, much of the hazardous waste generated in the United States was disposed in landfills or incinerators. When cement long kilns entered the picture in the late 1970's as a viable, even preferred option, a fierce competition was set in motion that continues to this day. The battle for waste was fought on every level – economic - with price wars, political - with lobbying efforts, and social - with a battle for public opinion. In the end, because of its inherent benefits, the cement kiln has emerged as the preferred option for most wastes. Of the 2 million tons of liquid hazardous waste generated in the United States in 1998, cement kilns captured a 62% market share. As regulations get tougher and technologies improve, the trend of cement kiln market dominance is expected to continue.

As one of the top three energy users in the world, the cement industry is clearly playing a leading role in the environmental reduction of waste through the use of waste as energy. With fossil energy costs increasing and profit margins decreasing, cement manufacturers around the world are using liquid waste fuels to cut costs, increase profits and breathe new life into their long cement kilns.

But liquid waste fuels, though important, are only the beginning. A little more than ten years ago, Cadence Environmental Energy recognised that many energy rich solid wastes could also be beneficially used by a cement kiln if a way could be created to introduce them into the kiln. The goal of using solid waste as cement kiln energy is clearly a prize worth striving for. Once accomplished, this new waste source could double a kiln's energy savings potential, and equally important, it would literally provide older long kilns with a new life by allowing them to compete with the more energy efficient preheater/precalciner kilns. Cadence teamed up with its technology partner Ash Grove Cement Company of Overland Park, Kansas, one of the country's leading cement manufacturers, to find a way to use these energy-rich solid wastes as fuel.

4. FEEDING WASTE ENERGY TO "HUNGRY" KILNS - WITH A FORK

The most modern kilns with the latest technology, the preheater/precalciner kilns, feed up to 60 percent of their fuel in the precalciner. This supplies the energy directly to the calcining zone, which is the most energy intensive part of the process. The joint development effort targeted the calcining zone as the proper place to add this supplemental fuel, even though, this zone is in the middle of a long kiln. This effort produced a technology many thought impossible - a device, called a Feed Fork, that could introduce solid wastes into the calcining zone of a rotating long kiln. While experts agree that the calcining zone is the proper place to add supplemental fuels, they were not certain that the Feed Fork could actually do it. Some voiced concerns that adding fuel to the middle of the long kiln might disturb the delicate fuel/energy balance needed to produce cement. But early tests proved that by making slight adjustments in the operating parameters and temperature profile of the kiln to account for the fuel at mid kiln, its thermal equilibrium could be reestablished with no negative impact on the process. Indeed, firing a portion of fuel in the calcining zone greatly stabilises the process and has a positive impact on the overall cement manufacturing process.

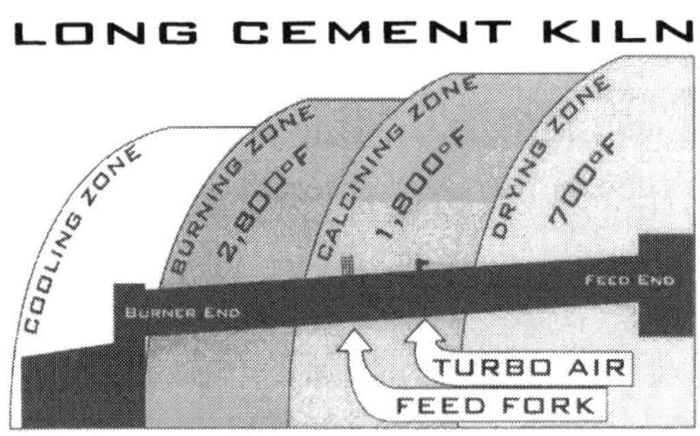

Figure 2. Long cement kiln indicating temperature zones and location of the Feed Fork and Mixing Air Turbo

The Feed Fork is dependable, tough, and mechanically simple, with only one moving part. It is also versatile enough to allow the use of many different waste types. Whole scrap tires, boxes, bales, bags, pails, if the waste can fit through the opening, chances are it can be used.

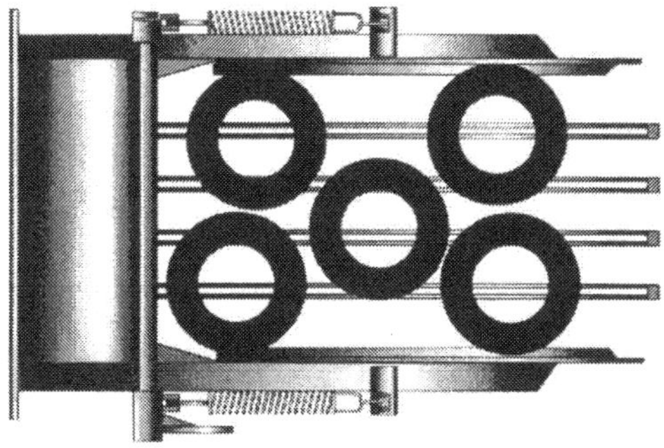

Figure 3. The Cadence Feed Fork showing scrap tires ready to be recycled as energy.

Since waste streams vary from location to location, the Feed Fork offers kiln operators substantial flexibility in using any waste that is abundant in their particular area, as well as being able to change from one waste fuel source to another quickly and easily.

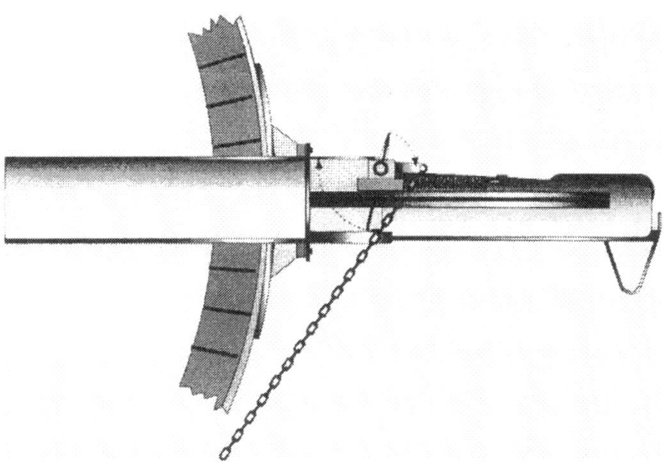

Figure 4. The Cadence Feed fork showing the drop tube penetrating through the kiln wall and the pivoting gate through which the waste enters

One of the most widely used sources of waste energy are whole scrap tires. Scrap tires are a perfect source of energy for long kilns. They are abundant, found in every industrialized country, pound-for-pound they contain more Btu value than coal, and is also a source of iron. The Feed Fork introduces whole tires into the kiln; no shredding or processing is necessary. Whole tires burn more slowly and evenly than tire chips and thus are a superior fuel than shredded tires.

Figure 5. US scrap tire market showing total volume and tires used for energy recovery.

In 1999 the number of scrap tires generated in the United States alone was 270 million. That is one scrap tire for every man, woman and child in the US. Each year the number grows by nearly 50 million scrap tires. Today, 64% of the scrap tires generated are used for energy recovery ... primarily in cement kilns. The reason is clear: scrap tire fuel generates revenue and saves fossil fuel costs.

A simple cost/benefit example tells the story.

ASSUMPTIONS:

- Coal $40/Ton - 12,000 Btus/#
- Tires 20# each - 15,000 Btus/#
- Resulting energy value per tire = $.50
- Kiln feed rate 823,000 Tires/year - $.30 per tire revenue to kiln

Revenue sayings potential to kiln		
	First year [$]	Each following year [$]
2,000,000 tires used as energy ($2.00/mmBTu) (20,000 tons)	1,000,000	1,000,000
Disposal revenue + $30/ton ($ 30/tire) disposal fee paid to cement company	450,000	450,000
Total	1,450,000	1.450,000

Initial costs		
Equipment/installation (2 kilns) Automated Conveyors & Feed Equipment	900,000	75,000
Permitting Fees	150,000	75,000
Total Costs	1,050,000	75,000
Ongoing annual operating costs		
Labor	120,000	120,000
Equipment	10,000	50,000
Maintenance	10,000	30,000
Total	140,000	200,000
Estimated annual net benefit	260,000	1,250,000

But fuel savings are only part of the story. Feed Fork users around the world have reported a positive impact on clinker mineralogy. The feed fork system creates a shorter more stable burn zone. Feed entering the burn zone has a quicker rise to temperature and a shorter residence time at sintering temperature. This improves the mineralogy and grindability of the clinker. Lower heat loading in the burn zone has also extended the life of the burn zone refractory. In addition to these benefits, there is a significant drop in the NO_x emissions.

The chart below shows the impact the Cadence Feed Fork has on NOx reduction. Six plants operating long cement kilns in the United States engaged in a test to show the impact of mid kiln fuel on NO_x emissions. The detailed analysis below proves that the addition of fuel at mid kiln can reduce NO_x emission by up to 50%.

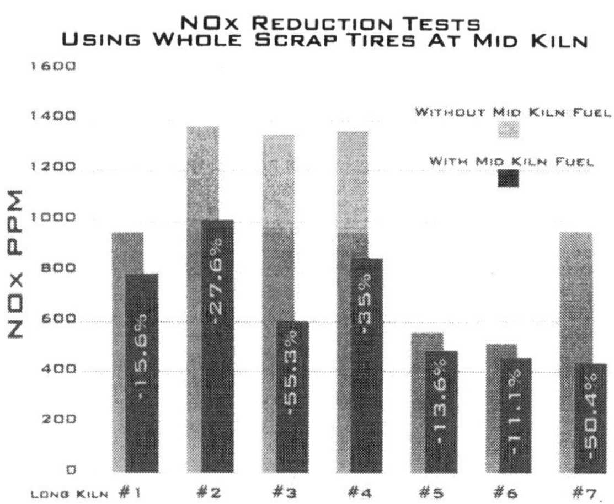

Figure 6. NO_x reduction comparison on six long cement kilns with mid kiln fuels and without mid kiln fuels

5. FEED FOR INSTALLATION AND PAYBACK

Installation of the Cadence Feed Fork can be accomplished during a typical maintenance shutdown period and investment pay back which is funded by fossil fuel savings and waste disposal fees, is typically one year or less.

When all the benefits are considered, it is easy to see why Cadence's Feed Fork technology has become a powerful profit-generating partner for long cement kiln operators around the world.

With nearly 15 years of dependable use, it is clear to see that waste as fuels are good for the profitability of the cement making business, good for the cement making process, and good for the environment since potentially beneficial wastes can be used rather than landfilled or incinerated.

6. BREAKING THE WASTE FUEL LIMITATION BARRIER

Depending on the process and kiln specific issues, a typical kiln can substitute waste for fossil fuels at a maximum rate of 15% to 25%. This maximum substitution rate is a technological barrier related to the kilns use of oxygen and flame temperature. But technological barriers can often be moved back with technological innovations. The fact is that a long cement kiln has the potential to more than double the amount of fuel used at mid kiln if the flame temperature could remain constant.

Again, the Cadence Ash Grove development team went to work searching for a way to supply air for the combustion of more waste fuel at mid kiln without increasing excess combustion air at the main flame. After 10 years of development, extensive computer modeling and testing, the puzzle was solved with a device called

'MAT', Mixing Air Turbo. While the technology is sophisticated, the idea behind MAT is fairly simple. Adding more fuel at mid kiln means adding more oxygen at mid kiln for proper combustion.

Numerous computer models designed to determine the dynamics of the airflow within the kiln and the combustion characteristics of the fuel used at mid kiln showed that mixing the air within the kiln could double the amount of waste fuels while enhancing the entire combustion process. The computer models showed significant layering of the combustion gasses inside the kiln. The Mixing Air Turbo takes air at high pressure and velocity and injects it into the kiln, imparting a rotational component into the airflow. By mixing combustion gasses and creating turbulence, NO_x and other emissions are significantly reduced while the amount of fuels used at mid kiln are increased.

When the Mixing Air Turbo was added, two significant benefits became instantly obvious.

1. The waste fuel substitution rate doubles.
2. Emissions are substantially reduced.

The Mixing Air Turbo was first installed and tested on the Ash Grove kiln in Chanute Kansas. From the very start it met and exceeded established expectations. Here are the results of the actual test showing the dramatic results.

These charts depict 56 hours of kiln operation. The kiln is operated with mid-kiln fuel for the first 32 hours. The Mixing Air Turbo is turned on and the effect on the kiln is recorded for the next 24 hours. Emission data was collected at one-minute intervals.

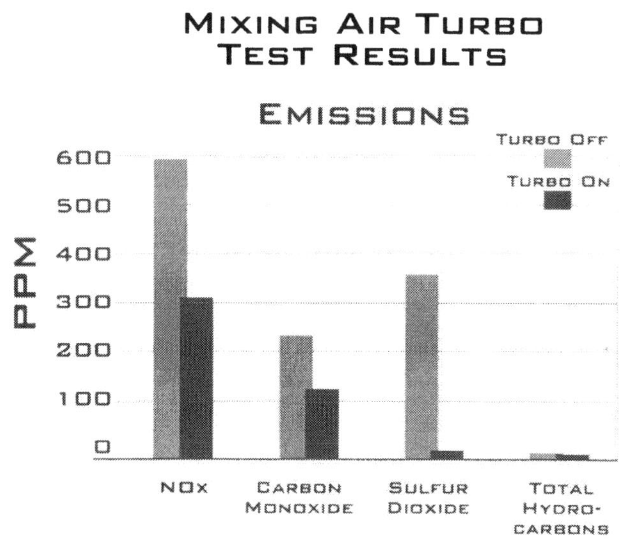

Figure 7. The effect of the Mixing Air Turbo on various emissions.

The facts speak for themselves. While using the Mixing Air Turbo there was almost a 50% reduction in CO, a 97% reduction in SO_2 Emissions, a 20% reduction in Total Hydrocarbon, and almost a 50% Reduction in NO_x.

During this test, the average solid waste substitution rate was more than doubled from 0.8 tons per hour to 1.8 tons per hour.

This type of increase in fuel substitution has been consistent in all of the testing Cadence has performed. The Mixing Air Turbo has had the predicted positive impact on the fuel substitution rate.

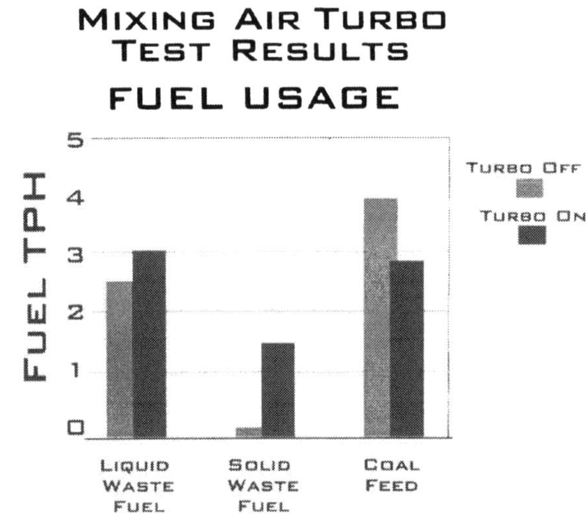

Figure 8. The ability of the Mixing Air Turbo to increase waste fuel input.

The theory behind these remarkable results is related to the flame temperature. When the Mixing Air Turbo is added to the kiln, the total airflow in the kiln is not increased. This means that under normal circumstances 100% of the air enters the kiln through the burn zone. With the Mixing Air Turbo about 10% of the air enters the kiln through the fan and 90% of the air now comes through the burn zone. This decrease in volume of the secondary combustion air increases the temperature of the flame. The fuel to air ratio approaches what it was prior to the addition of fuel mid-kiln. This increase in flame temperature stabilizes the kiln and allows for additional fuel substitution.

When Cadence engineers reviewed the operation of the mid-kiln and Mixing Air Turbo system they were surprised to find that it incorporated the latest NO_x reduction technology. The mid-kiln fuel creates a secondary combustion zone at a lower temperature. This secondary combustion zone, as well as the extra oxidizing potential of the Mixing Air Turbo, creates a staged combustion system within the kiln which has been recognized as the best technology for NO_x reduction.

Using solid waste as energy in long cement kiln is a technology that is available to every industrialized country - rich or poor, right now. It transforms long cement kilns into waste burning, revenue-generating regional assets that build the infrastructure while cleaning the air and protecting the environment.

The complete mid kiln technology package from Cadence, including the Feed Fork and Mixing Air Turbo is a real system, providing real benefits on real kilns around the world today.

To learn more about our technology or to contact Cadence, visit our web site at www.cadencerecycling.com.

THE PYROLYTIC CONVECTIVE WASTE UTILISER

Tadeusz Piecuch
Departament of Water, Wastewater and Waste Technologies, Koszalin University of Technology, 15-17, Racławicka Strr., 75-620 Koszalin, Poland

Key words: pyrolisis, waste utiliser

Abstract: Described here is the construction and principles of action of the utiliser and the influence of its operations on air pollution. The paper presents the results of the author's own research on the degree of leachability of pollutants from the solid secondary wastes left after such pyrolysis.

1. CONSTRUCTION AND PRINCIPLES OF ACTION

The pyrolytic convective waste utiliser (hereinafter the WPS) is a device operating periodically and serving in the thermal transformation of wastes using electricity[1]. It is produced in three versions with working volumes of the chamber equalling 350, 1000 and 2500 dm³. The complete WPS installation as presented in Figure 1 comprises three sub-systems: the pyrolytic chamber, the unit filtering gases and the burner in which the post-pyrolytic gas is burnt.

Wastes designated for disposal via the WPS are introduced into the pyrolytic chamber (2) shown in Figure 2. The wastes may be introduced loose or in packaging (such as plastic sacks or barrels, cartons, cardboard boxes, etc.). Following the loading of the chamber (2) and its closure, wastes are subjected to thermal processing with the aid of a heating unit (11). The process of pyrolysis most often takes place at a temperature of 600-750°C. The pyrolytic gas arising at this point rises by laminar motion and makes contact with the calcic sorbents with which the charge is covered (14). These constitute the zone of preliminary filtration of the pyrolytic gas, which is then directed to the so-called zone of gas overheating (15). Here the prevailing temperature is somewhat higher, in the range 750 to 850°C. After passing through this, gas is usually directed straight to the burner - Figure 1C. The filtration unit shown in Figure 1B is mostly used rather rarely, since the need for it does not often arise.

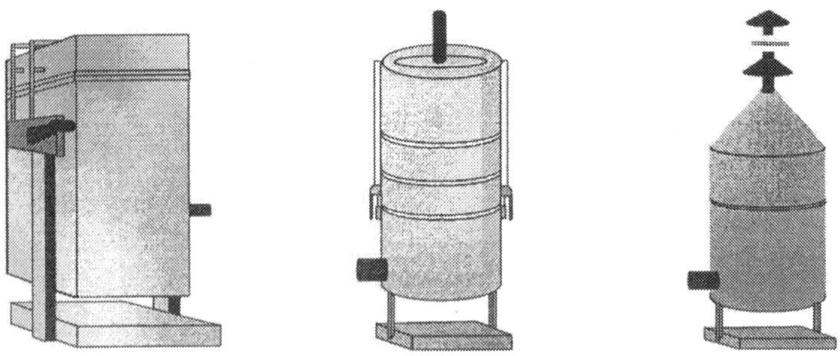

A B C

Figure 1. Utiliser main sub-assemblies: A – pyrolytic chamber, B – filtration assembly, C – burner[26].

The characteristic feature of the operation of the WPS utiliser is the addition "from above" - to the already-deposited layer of wastes - of the calcic sorbent which brings about the binding-up of sulphur (as $CaSO_4$), and chlorine (as CaCl2), within the so-called fly ash, i.e. the residue following processing. The proper course of the process requires a high degree of moistening of the processed waste. If it is dry, then water must be added. At the beginning of the process, the pressure of the water vapour drives off air from the air spaces of the wastes found in the utiliser before the wastes begin to undergo thermal breakdown. This driving-off of air from the air spaces in the wastes ensures that there is no oxygen in the centre of the utiliser during pyrolysis.

The construction of the vessel within the filtration-sorption unit is shown in Fig. 3. This is a type of separator comprising three main elements; the closing cover (1), the carbon-limestone filtration-sorption bed (15) and the gas scrubbers (5).

Post-pyrolytic gas enters the pyrolytic chamber (2) - Figure 2, is directed to the vessel of the filtration-sorption array - Figure 3 via a conduit (6), and then flows into the gas condenser (4) in which condensate is formed within the space (5). This part of the vessel of the filtering unit is known as the gas scrubber (5). Condensate falls by gravity onto the carbon filtro-sorbent and is then discharged externally.

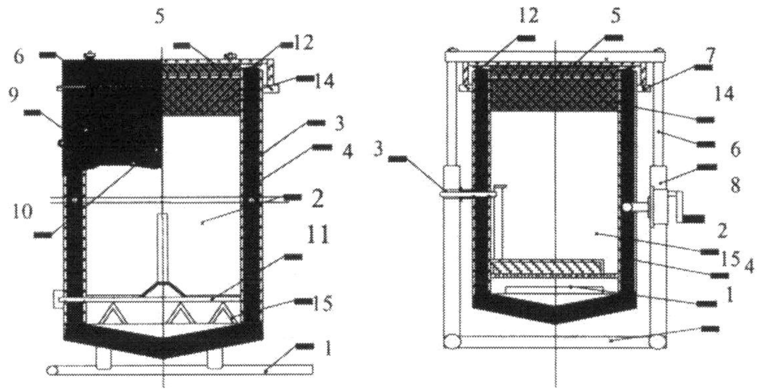

Figure 2. Pyrolytic chamber: 1 – supports, 2 – chamber, 3 – outer housing made of stainless steel., 4 – thermal insulation, 5 – cover, 6 – telescope, 7 – clamps, 8 – bearing, 9 – cover lift gears, 10 – pyrolytic chamber rotation gears, 11 – heating elements assembly, 12 – gaskets, 13 – gas outlet, 14 – initial filtration zone, 15 – gas superheating zone.

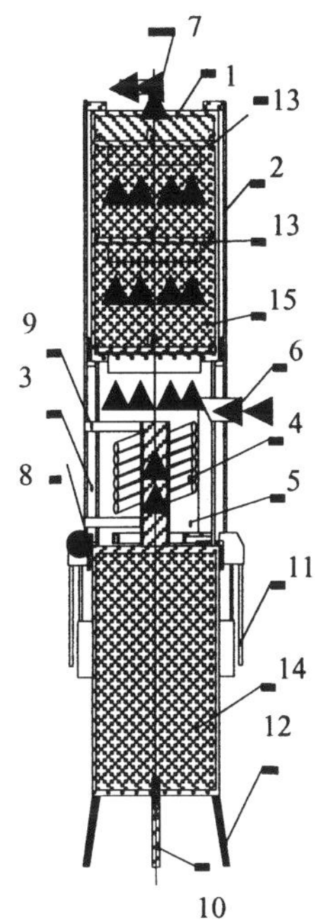

Figure 3. Filtration assembly: 1 – cover, 2 – filter housing, 3 – cooler housing, 4 – gas cooler, 5 – gas washer, 6 – gas inlet from pyrolytic chamber (see Fig. 2.), 7 – gas outlet to burner (see Fig. 4.), 8 – cooling water inlet, 9 – cooling water outlet, 10 – gas washer drain valve, 11 – closing clamps, 12 – supports, 13 – gaskets, 14 – reflux carbon filter, 15 – filtration-sorption bed.

In contrast, the post-pyrolytic gas, having passed through the condenser, moves up the vessel of the filtration-sorption array and passes over the sorption beds before leaving the chamber via the outlet pipe (7) and being directed to the burner - Figure 1C.

The construction diagram for the burner used here is shown in Figure 4.

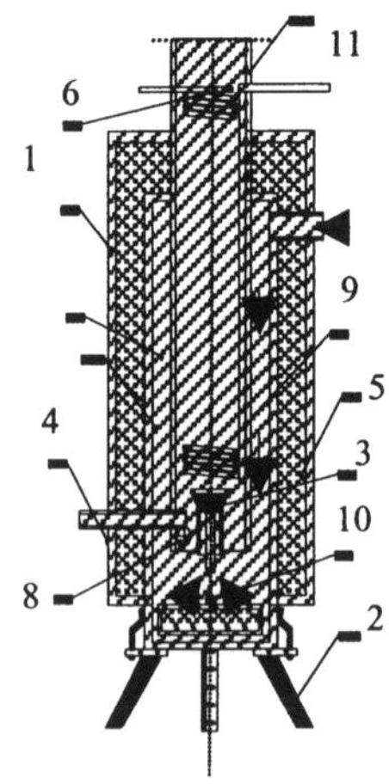

Figure 4. Burner[26]: 1 – housing, 2 – supports, 3 – burner, 4 – gas igniter, 5 – thermal insulation, 6 – thrust breaker, 7 – gas superheating chamber, 8 – fan, 9 – heat exchanger, 10 – water drain valve, 11 – chimney.

In the burner, the main combustion chamber (1) - Fig. 4 comprises the gas burner proper (3) and the gas-heating chamber (7). The construction of the burner allows for precise burning of the generated post-pyrolytic gas - irrespective of changes in the amounts involved. This is assured by the in-built ventilator (8), as well as the wattles resistance wire (9). This wire becomes incandescent, bringing about a stable oscillation of the gas in the burner chamber and hence stabilising the combustion and giving the lack of any visible flame that attests to full combustion of the gas. To increase the efficiency of combustion an automatic valve is mounted within the burner, directly on the intake nozzle to the post-pyrolytic gas burner. The valve provides for the accurate portioning of air and the gas in the ratio 14:1 by volume. This ensures the constant activation of the burner. The automatic valve introducing an appropriate amount of air into the burner is founded in a perforated cylindrical partition (the perforation of the partition is in the form of apertures with quasi-tangential pipes to the external cylindrical surface). This gives rise to a rotational movement of air at the point of entry into the burner (3) and hence provides for excellent mixing of the gas with air and therefore very good combustion.

Also important for the quality of the combustion is the maintenance of proper proportions and volume in the case of the burner itself - Figure 4.

The combustion temperature must be selected optimally, i.e. high enough to ensure combustion of heavier hydrocarbons and their derivatives, but also low enough to minimise the generation of oxides of nitrogen[7,8,9,10,11]. Depending on the composition of the gas, the temperature may range across rather wide limits.

2. RESEARCH RESULTS

2.1 The Influence of Operations of the Pyrolytic Waste Utiliser on the Pollution of the Air – Composition of Waste Gases Generated from Post-pyrolytic Gas

Research was first done using the processing of hospital wastes as an example. The charge of 52 kg comprised 9 kg of textile wastes, 8 kg of organic tissue, 8 kg of paper, 20 kg of plastics and rubber, 4 kg of garbage, 0.5 kg of oudated medicaments and 1.5 kg of glass and metal. The post-pyrolytic gas obtained from this contained:

Post-pyrolytic-gas	[%]
Hydrogen - H_2	10.26
Oxygen - O_2	0.62
Nitrogen - N_2	3.26
Carbon monoxide - CO	1.60
Carbon dioxide - CO_2	6.80
Methane - CH_4	20.30
Ethene - C_2H_4	11.10
Ethane - C_2H_6	10.00
Propene - C_3H_6	12.60
Butene - C_4H_8	0.09
i-butene - C_4H_8	0.33
i-butane - C_4H_{10}	5.80
Butane - C_4H_{10}	0.18
i-pentene - C_5H_{10}	0.87
pentene - C_5H_{10}	3.88
i-pentane - C_5H_{12}	3.80
hexane - C_6H_{12}	0.06
i-hexane - C_6H_{12}	2.18
heptene - C_7H_{14}	3.26
i-heptene - C_7H_{14}	0.05
i-heptane - C_7H_{16}	0.25
benzene - C_6H_6	2.50
i-octane - C_8H_{18}	0.99
toluene - C_7H_8	1.10
	101.88%

Table 1 (A, B) presents the results of analytical determinations of concentrations (one-off in the course of 30 minutes) in the waste gases arising during the processing of the mentioned wastes (column 2), as well as - for comparison - during the processing of phthalo-carbamide paint wastes (column 1), white chemo-hardenable enamel with hardener (column) and wastes of polyester resins containing maleic anhydride, glycol and styrene (column 4)[3].

Table 1A. Concentrations determined once in the course of 30 minutes in the waste gases from different kinds of waste subjected to pyrolysis, and maximum permissible D30 concentrations denoted in the course of experiments at the landfill at Ostrów Grabowski near Szczecin [μg/m³][9,10].

Type and quantity of waste	SO_2		CO_2		NO_2	
	perm.	noted	perm.	noted	perm.	noted
Phthalo-carbamide paints charge – 161 kg, gas throughput – 102 m³/h	300.0	116.0	2,500.0	9.0	250.0	4.0
Hospital wastes charge – 52 kg, , gas throughput – 98 m³/h	300.0	4.0	2,500.0	122.0	250.0	2.0
White chemohardenable enamel with hardener charge – 220 kg, , gas throughput – 43 m³/h	300.0	3.0	2,500.0	12.0	250.0	2.0
Polyester waste – maleic anhydride, glycol and styrene carge – 43 kg; , gas throughput – 203 m³/h	300.0	35.0	2,500.0	354.0	250.0	14.0

Table 1B. Concentrations determined once in the course of 30 minutes in the waste gases from different kinds of waste subjected to pyrolysis, and maximum permissible D30 concentrations denoted in the course of experiments at the landfill at Ostrów Grabowski near Szczecin [μg/m³][9,10].

Type and quantity of waste	HCl		Cl_2		PAHs	
	perm.	noted	perm.	noted	perm.	noted
Phthalo-carbamide paints charge – 161 kg, gas throughput – 102 m³/h	80.0	1.5	40.0	0.6	800.0	4.0
Hospital wastes charge – 52 kg, , gas throughput – 98 m³/h	80.0	12.0	40.0	0.6	800.0	71.0
White chemohardenable enamel with hardener charge – 220 kg, , gas throughput – 43 m³/h	80.0	1.0	40.0	0.6	800.0	15.0

Type and quantity of waste	HCl		Cl_2		PAHs	
	perm.	noted	perm.	noted	perm.	noted
Polyester waste – maleic anhydride, glycol and styrene carge – 43 kg; , gas throughput – 203 m^3/h	80.0	nd.	40.0	-	800.0	4.0

The concentrations of pollutants determined for the waste gases arising from the processing of the different kinds of waste were compared with D30 concentrations (i.e. those permissible in 30-minute determinations), and were found to be between 1 and 2 orders of magnitude lower than the permissible maxima, i.e. completely safe.

Table 2 (A, B) gives the permissible emissions for the pollutants considered, as laid down by the State Environmental Protection Inspectorate in Gdansk, which is to say the body to which the WPS utiliser tested is subordinated, being located at the Waste Utilisation Plant on the landfill in Szadółki near Gdansk[4].

Table 2A. Concentrations determined once in the course of 30 minutes in the wastes gases from phthalo-carbamide paint and carbamide-formaldehyde resin wastes subjected to pyrolysis at the site in Szadółki near Gdańsk [$\mu g/m^3$][9,10].

Type and quantity of waste	SO_2		CO_2		NO_2	
	*perm.	noted	perm.	noted	perm.	noted
Phthalo-carbamide paint, carbamide-formaldehyde waste charge – 160 kg, , gas throughput – 102 m^3/h	300.0	118.1	2,500.0	4.32	250.0	9.81

* denotes permissible normative D30 values.

Table 2B. Concentrations determined once in the course of 30 minutes in the wastes gases from phthalo-carbamide paint and carbamide-formaldehyde resin wastes subjected to pyrolysis at the site in Szadółki near Gdańsk [$\mu g/m^3$][9,10].

Type and quantity of waste	HCl		Cl_2		PAHs	
	perm.	noted	perm.	noted	perm.	noted
Phthalo-carbamide paint, carbamide-formaldehyde waste charge – 160 kg, , gas throughput – 102 m^3/h	60.0	14.13	60.0	3.14	1,200.0	4.32

2.2 Test on the Leachability of Pollutants from the Secondary Waste Constituted by the Post-pyrolytic Residue

The residue generated following the pyrolysis of wastes and constituting a secondary waste took the form of a fly-ash with about 4-8% of the mass of the original charge[5,6,7]. The possibility of dumping such residue at a landfill site depends on the degree to which pollutants may be leached from it and subsequently find their way into groundwaters and reservoirs. Experimental measurements of leaching were therefore carried out using a glass column of 0.05 m diameter containing a layer of secondary waste of 0.981 dm^3 settled volume, through which distilled water was run. The volume of water corresponded to a column of 760 mm, a quantity close to the amount of rainfall received in Poland's Baltic coastal belt. The results presented in Table 3 concern 16 different secondary wastes tested in the above manner:

1 - residue from the processing of PET bottles and sawdust, as well as other woody wastes accounting for c. 50% by volume;
2 - residue following the liquidation of acrylic and phthalic paints, again in the proportion 1:1 by volume;
3 - residue from the liquidation of acrylic paint wastes;
4 - residue from the liquidation of phthalic paint wastes;
5 - residue from the liquidation of PET bottles;
6 - residue from the liquidation of plastic wastes and leather in the form of old footwear;
7 - residue from PET packagings and outdated cosmetics (creams, oils, soaps and shampoos) in the ratio 1:1 by volume;
8 - residue from the liquidation of wastes as in 6 - another manufactured unit;
9 - residue from the liquidation of plastics and cosmetics in the ratio 2:1 by volume;
10 - residue from the liquidation of old tyres;
11 - residue from the liquidation of wastes - PET bottles and sawdust in the ratio 2:1 by volume;
12 - residue from the liquidation of tar wastes;
13 - residue from the liquidation of tar with sawdust at proportions of 1:1 by volume in the charge;
14 - residue from the liquidation of mineral-fibre wastes and sawdust in the ratio 1:1 by volume;
15 - residue from the liquidation of different pesticides and sawdust in the ratio of 1:4 by volume respectively;
16 - residue from the liquidation of pesticides.

In cases in which the type of processed plastic is given as PET, there was no PVC present. In the cases of other plastics the additional presence of polyvinyl chloride can not be ruled out.

The results contained in Table 3 indicate that the leachates from the different post-pyrolytic residues differed markedly from one another as regards chemical oxygen demand (COD) and dissolved matter content. As is seen from data in Table 3 and elsewhere[4,5,6,7], the basic components of the

Table 3. Results of determinations of pollutant concentrations in the eluate from secondary post-pyrolytic wastes obtained from different kinds of primary wastes[14].

Exp. No.	pH	BOD	COD	St	Sm	Dr	Ee	Ph
1	10.4	nd	333	110	3,500	3,610	nd	nd
2	8.4	nd	402	86	2,800	2,886	nd	nd
3	4.55	18	744	36	171	207	35	ND
4	7.35	14	663	140	170	310	18	ND
5	9.0	10	456	216	484	700	24	ND
6	10.6	4.0	22,327	830	14,072	14,903	171	ND
7	12.6	4.0	16,657	2,440	21,735	24,175	238	ND
8	10.6	5.0	44,651	909	13,874	14,783	179	nd
9	12.5	7.0	33,317	2,565	21,149	23,714	231	nd
10	10.7	9.0	1,318	556	6,209	6,765	848	nd
11	10.4	5.0	7,235	1,237	21,552	22,789	179	nd
12	11.6	231	46,032	12,768	31,418	44,186	389	nd
13	4.7	19	5,159	404	2,214	2,618	199	nd
14	4.4	2.0	1,290	174	1,174	1,248	115	nd
15	9.6	202	2,560	1,224	1,514	2,738	50	nd
16	6.0	6	646	2,731	20,975	23,706	949	nd
Standard	6.5-9.0	30	150	50	2,000	-	50	0.5

BOD [mg O_2/dm^3]
COD [mg O_2/dm^3]
St – total suspended matter [mg/dm^3]
Sm – soluble matter [mg/dm^3]
Dr – dry residue [mg/dm^3]
Ee – ether extract [mg/dm^3]
Ph – volatile phenols [mg/dm^3]
nd – not determined
ND – not detected

eluates are mineral chlorides and sulphates and a small amount of phenol-like organic matter. There are relatively high concentrations of sodium and calcium, which may derive from the high inputs of physiological solutions, etc., in the case of hospital wastes. Trace amounts of chromium were also noted in the eluates. The results of the research presented in Table 3 and of other studies in the literature[8,9,10,11,12,13,14] suggest that the post-pyrolytic residues in the form of "fly-ash" cannot be dumped directly into landfills, but rather require additional processing to reduce the leachability of pollutants (e.g. solidification using binders). Research into such possibilities is being continued at the Department of Water, Wastewater and Waste Technologies at Koszalin University of Technology. Attempts at manufacturing pavement blocks and slabs from them are also being made[15,16,17].

3. CONCLUSION

1. Pyrolysis in the WPS installation may be a technically, economically and environmentally suitable solution when it comes to the management of hospital, paint and lacquer wastes, as well as plastics, cosmetics and rubber goods.
2. The solution described is not however a favourable one where the management of waste coal-tar, pesticides and chemical warfare agents is concerned.
3. A certain technical and economic problem is posed by the secondary waste arising following the pyrolysis of wastes in the WPS, since this needs to be subjected to further processing, e.g. by solidification.
4. Pyrolysis in WPS installations allows for the sanitisation of wastes and for a marked reduction in their mass and volume, without giving rise to air pollution and with a possibility for a certain amount of energy to be gained from the burning of post-pyrolytic gas.
5. Simplicity of construction and operation predestine WPS installations for use in the management of wastes in small localities where the construction of incinerators is not justified economically or technically on account of the small amount of waste generated.

REFERENCES

1. Piotrowski, J., Wisniewski, W., Konwekcyjny utylizator odpadów medycznych i przemysłowych, Patent No 171673 i No 173648.
2. Namieśnik, J., 1999, Pomiary emisji zanieczyszczeń do atmosfery ze spalania gazów odlotowych z prototypowego urzadzenia do pirolitycznego rozkładu odpadów szpitalnych, Zakład usług OMNISFERA. Gdańsk.
3. Wyniki pomiarów kontrolnych emisji gazów na stanowisku Pirolitycznego Konwekcyjnego utylizatora Odpadów Typu WPS-360 wyposażonego w palnik pyłowy w komorze dopalania SP-50. PIOŚ-WIOŚ – pomiary grudzień 1994 – grudzień 1996. Szczecin.
4. WOŚ UW – Gdańsk – decyzja z dnia 26.10.1996 dotycząca dopuszczalnej emisji zanieczyszczeń z Utlizatora Odpadów typu WPS-350 zainstalowanego na wysypisku śmieci w Dzadółkach.
5. Szymański, J., 1996, Ocena oddziaływania na środowisko zastosowania dwóch utylizatorów pirolitycznych typu WPS-1000 w ZOZ w Ostrowii Mazowieckiej (ul. Dubois 68, Warszawa).
6. Skalmowski, K., 1994, Badania wyciągów wodnych koksu po procesie wytlewania odpadów medycznych oraz bloczków betonowych powstałych w wyniku cementacji koksu z odpadów medycznych. ISIŚPW. Warszawa
7. Merzyńska, A., and Przepiery, A., 1994, Badania wyciągów wodnych i bloczków betonowych. CSChN PAN. Police.

8. Grochowalski, A., 1997, Badania zawartości substancji toksycznych w gazach spalinowych ze spalarni odpadów niebezpiecznych i przemysłowych. III Międzynarodowa Konferencja *Spalanie odpadów – technologie i problemy*. Szczyrk
9. Grochowalski, A., 1998, PCDD and PCDF Cncentration in Gases and Bottom Ash from Incineration of Hospital Wastes in Poland. *Chemosphere* No 37/9-12.
10. Grochowalski, A., 1999, Problemy analityczne oznaczania polichlorowanyc dibenzodioksyn i dibenzofuranów. IV Ogólnopolska Konferencja naukowa *Kompleksowe i szczegółowe problemy Inżynierii środowiska*. Koszalin-Ustronie Morskie.
11. Sokołowski. M., 1994, Dioksyny – ocena zagrożeń środowiska naturalnego oraz metody ich wykrywania. WIChIR, PIOŚ WARINTECH. Warszawa.
12. Edited by Pawłowski L., 1997, Utylizacja odpadów niebezpiecznych w piecach cementowych. PL Lublin: pp. 164.
13. Leszczyński, S., 1996, Piroliza gleb skażonych substancjami ropopochodnymi. Konferencja Naukowo-Techniczna Unieszkodliwianie i utylizacja odpadów niebezpiecznych. AGH Kraków.
14. Leszczyński, S., Piroliza odpadów celulozowych. IV Ogólnopolska Konferencja Naukowa Komplekowe i szczegółowe problemy inżynierii środowiska. Koszalin-Ustronie Morskie. WbiIŚ PK – *Inżynieria Środowiska*. Koszalin.
15. Piecuch, T., Dąbrowski, T., Hryniewicz, T., and Żuchowicki, W., 1998, Polish-Made Pyrolytic Convetive Waste Utiliser of Type WPS. Part 2. Problems of Residue Management after Thermal Waste Utilisation. 14th International Conference on *Solid Waste Technology and Management*. University of Pennsylvania and Widener University – Chester. Philadephia.
16. Piecuch, T., Dąbrowski, T., Hryniewicz, T., and Żuchowicki, W., 1998, Utilisation and Neutralisation Studies of Secondary Wastes from the Polish-Made Pyrolytic Convective Waste WPS-Type Utiliser. 14th International Conference on *Solid Waste Technology and Management*. University of Pennsylvania and Widener University – Chester. Philadephia.

RESEARCH ON POSSIBILITIES OF SEWAGE SLUDGE UTILISATION IN THE PROCESS OF PYROLYSIS

January B. Bień, Jurand D. Bień and Beata Matysiak
Technical University of Czestochowa, Faculty of Environmental Protection and Engineering, Dabrowskiego 69, 42-200 Czestochowa, E-mail: jbien@is.pcz.czest.pl

Key words: sewage sludge, thermal utilisation, pyrolysis, char, oil, gas

Abstract: Article presents the results of investigations on the pyrolytic process of utilisation of sewage sludge originated from the municipal-industrial sewage treatment plant. The products received during the process of the pyrolysis were subject to the detailed qualitative and quantitative analyses as far as their energetic usefulness is concerned.

1. INTRODUCTION

The most common method of sewage sludge thermal utilisation is combustion. However, high investment and exploitation costs induce searching for new technologies that should be as efficient as combustion but less expensive. At present in Poland the thermal method of sewage sludge utilisation is rarely used, but it does not mean that this situation will be still the case in the nearest future. Nowadays there is a number of new methods of thermal sludge utilisation around the world, either being in a development stage or already implemented. Among them we can find the following[1]:

- technology for sludge drying, called "Bojner"
- technology for thermal waste decomposition, called "Krepro"
- technology of wet oxidation.

In the recent years there has been a great interest in pyrolytic processes among researchers dealing with waste utilisation. Such processes carried out in a relatively low range of temperatures, allow for obtaining from waste such energetic products as: fuel gas as well as liquid and solid postpyrolytic products. These processes are carried out in non-oxygen environment, in a temperature range between 300 -700°C, and because of their endothermic character, they are much easier to be kept under control[2]. The temperature in which the decomposition of organic substances is taking place plays a crucial role in the whole process. Both, a efficiency of the pyrolysis as well as percentage of particular postpyrolytic products received during the process depend on that temperature. Barański[3] differentiated several processes that occur in certain temperatures, and they are as follows:

250°C	separating water and CO_2, depolymerisation, first phase of separating sulphur compounds
340°C	decomposition of aliphatic compounds, first phase of separating methane and other gas hydrocarbons
380°C	low temperature gas carbonisation
400°C	decomposition of organic acids as well as organic nitrogen compounds
500°C	alteration of bituminous substances into low temperature carbonised oil or pitch (tar)
600°C	cracking of oil into short chain gas hydrocarbons and formation of aromatic compounds
> 600°C	olefin dimerisation and formation of high boiling temperature aromatic compounds

Specific characteristics of pyrolysis make it useful for neutralisation of waste with large amount of components, containing such non-organic substances as: NaCl, $FeCl_2$, Zn, Pb as well as high level of Cl, S and/or N[2]. All of the mentioned features led to the application of pyrolysis for municipal and for dangerous industrial waste as it has been described by several publications. For example research conducted by Sobolewski and others[4] gave evidence that this method might be useful for utilisation of single-use medical equipment. Ammann[5] was carrying out tests on utilisation of dry batteries and accumulators, Pinto[6] investigated a decomposition process of wrappage made of plastics and aluminium. Tesch and Moergeli[7] were analysing the process of varnish waste utilisation. A lot of attention - as far as the usage of pyrolysis is concerned - was also devoted to utilisation of rubber waste, especially used tyres and transporter belts. One may assume from those examples that many similar experiments were also carried out on possibilities of sewage sludge utilisation by means of pyrolytic processes. However the authors have found only a few papers and firms capable of dealing with this problem. This can be a proof that the described method is still not popular. The most often used technology, which has been already tested in pilot installations was OFS method (Oil From Sludge)[2]. This technology was described as a low-temperature conversion of sewage sludge into liquid oils and char, i.e. into products that might be used as fuel. In order to assess a viability of the pyrolysis for sewage sludge utilisation we decided to carry out our own investigation.

2. EXPERIMENTS

Thermal waste utilisation was carried out on the testing system shown in Figure 1.

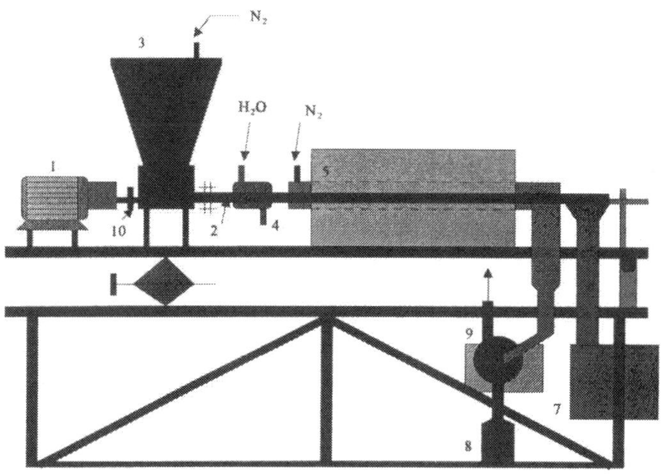

Figure 1. Diagram of the testing system:
1 - engine with moto-reducer, 2 - helical feeder, 3 - resources container, 4 - radiator, 5 - furnace, 6 - glass radiator, 7 - char container, 8- oil container, 9 - gas offtake, 10 - clutch

The stand comprises a reactor made of stainless steel with the internal diameter of 55 mm and the length of 1200 mm. The reactor is located inside the 3-heating zone furnace (5). Dried sludge is located inside the container (2), and then inserted into the reactor by means of the helical feeder. The time needed for the whole process is regulated by the engine's gearbox (1). Products generated during the pyrolysis are received in two containers: the char container (7), the oil container (8) and gas is taken with the use of the gas sample (9). For the research purposes raw sewage sludge from preliminary sedimentation tanks of Częstochowa municipal-industrial sewage treatment plant was used. Table 1 presents an analysis of sludge used for the test.

Table 1. Characteristic of raw sludge from the Częstochowa sewage treatment plant

Specification	Unit	Value
Colour	-	Brown (russet)
Smell	-	Org. specific
Reaction, pH	-	7.4
Dry residue	g/m^e	6.6
Hydration	%	93.4
Content of organic substances	%	71.0
Content of mineral substances	%	29.0
Heavy metals	mg/kg s.m.	
Ni		165
Cr		42
Pb		130
Zn		2175
Cu		175
Cd		11

The following four values of temperatures were chosen for carrying out the experiments: 350°C, 450°C, 550°C and 650°C. Those temperatures were used for conducting a thermal process of waste destruction generating different solid, liquid and gas products. Each time 500 g of dried sludge was inserted into the reactor. The duration of each process was 1,5 hours. The tests included the following steps:

- quantitative analysis of products received in the process of sewage sludge pyrolysis,
- qualitative analysis carried out to determine calorific values of received products.

3. RESULTS

The impact of temperature on generation rate of different products during the process of sludge pyrolysis is shown in Figures 2, 3 and 4, respectively.

The tests showed that the generation rate of char decreased with temperature, whereas generation rates of oils and gases increased with temperature. The determined in different temperatures fractions of oil, char and gas were tested to determine their heats of combustion and calorific values. The results are presented in Tables 2, 3 and 4 for oil, char and gas respectively.

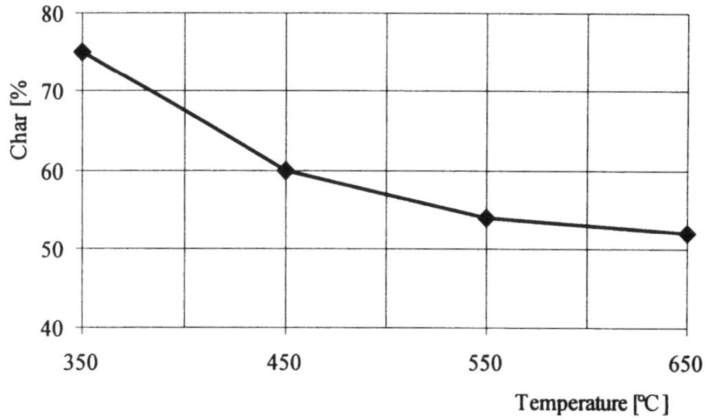

Figure 2. Rate of char generation during the sludge pyrolysis.

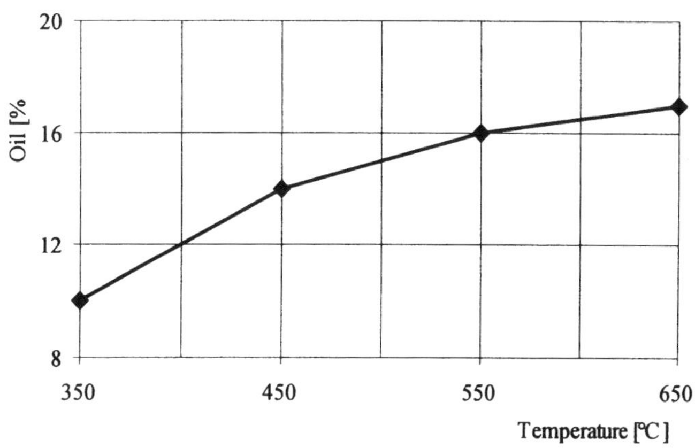

Figure 3. Rate of oil generation during the sludge pyrolysis.

Table 2. Calorific value and heat of combustion of oil received in different temperatures during the process of pyrolysis.

	Unit	350°C	450°C	550°C	650°C
Heat of combustion	kJ/kg	25506	27335	26127	26067
Calorific value	kJ/kg	23878	26039	25146	24934

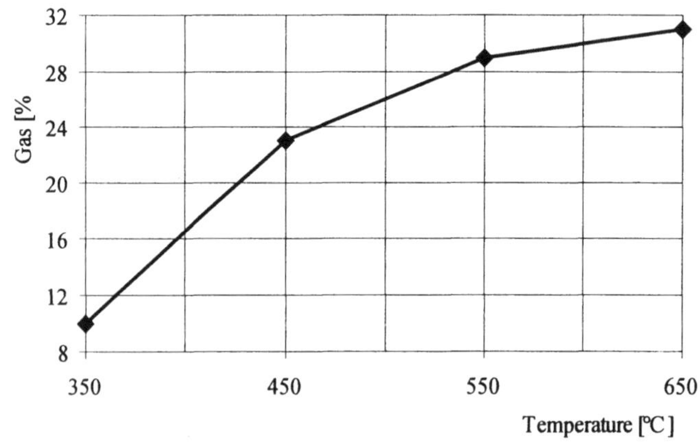

Figure 4. Rate of gas generation during the sludge pyrolysis

Table 3. Heating value and heat of combustion of char received in different temperatures during the process of pyrolysis.

	Unit	350°C	450°C	550°C	650°C
Heat of combustion	kJ/kg	11620	11590	10690	10360
Calorific value	kJ/kg	10998	11020	10135	988

Table 4. Heating value of gas received in different temperatures during the process of pyrolysis..

	Unit	350°C	450°C	550°C	650°C
Calorific value	kJ/m^3	5247	5308	5312	5398

The test results qualify the particular products as fuel or as an addition to other fuels. The calorific value of oil reaches the level of 27 MJ/kg, which is in typical for fuel oils range of 20-40 MJ/kg[9].

The next phase of investigations was carried out to characterise qualitatively and quantitatively chemical compounds in the process of pyrolysis. A tendency of heavy metal accumulation in the char was observed. Significant concentrations of cadmium, zinc and lead were detected in the pyrolytic oil (refer to Table 5)[10].

In additional test the postpyrolytic oil and gas received in temperature of 450°C were subjected to a chromatographic analysis. 29 different compounds were identified in the oil. The identified compounds and their amounts are shown in Tables 6 and 7, respectively.

Table 5. Content of heavy metals in the postpyrolytic products[9].

Metals	Concentration [mg/kg s.m.]	
	Char	Condensate
Nickel (Ni)	133.8	32.0
Chromium (Cr)	421.5	9.3
Lead (Pb)	74.15	55.7
Zink (Zn)	1311.3	863.5
Cooper (Cu)	160.4	14.5
Cadmium (Cd)	0.68	10.3

Table 6. The compounds identified in the pyrolytic gas.

	Name of the compound	Concentration [mg/m³]
1.	CO	995
2.	SO_2	24.8
3.	NO_X	423
4.	HF	16.3
5.	HCl	416.8
6.	H_2S	198.7
7.	NH_3	> 1150
8.	Benzene	69.95
9.	Toluene	1004.62
10.	Ethylobenzene	261.29
11.	Xylene	53.07
12.	Benzαopiren	1.78

Table 7. PAH compounds and their amounts detected in oil.

	Name of the compound	Concentration [mg/g]
1.	toluene	0.62
2.	pyridine	0.092
3.	3-methylo-1-cyclopentene*	-
4.	2-methylpyridine	0.26
5.	m-, p-xylene, ethylobenzene	0.70
6.	o-xylene	0.32
7.	styrene	1.50
8.	n-methylo-pyridine-2-amine	0.16
9.	2,4-two-methylpyridine	0.23
10.	phenol	1.56
11.	benzofurane*	-
12.	Benzonitryle*	-
13.	4-methylo-ethynylobenzene	0.73
14.	o-cresol	0.34
15.	m.-, p-cresol	1.35
16.	4-methylobenzenoisocyanin*	-
17.	7-methylobenzofurane*	-
18.	1,2-two-hydronaphtalene*	-
19.	2(1H)-chinoline	0.27

	Name of the compound	Concentration [mg/g]
20.	naphtalene	0.97
21.	isochinoline	0.25
22.	1-methylonaphtalene	0.42
23.	2-methylonaphtalene	0.51
24.	benzylonitryle	0.70
25.	1,2-two-methylonaphtalene*	-
26.	4-methylo-1H-indole*	-
27.	1,2-two-hydroacenaphtalene	-
28.	9H-fluorene	-
29.	anthracene	0.13

* concentration determination was subject to significant analytical error

The above results show that condensed hydrocarbons occur always in situation when organic substances are exposed to high temperature. The high concentration of pollutant in gaseous products indicates a necessity of introducing purification processes in pyrolytic installations used for sewage sludge utilisation.

4. CONCLUSIONS

The obtained results show that the pyrolysis process may find practical applications but its economics depend on further development of technologies allowing for utilisation of the oil as well as heat recovery from char combustion. The flue gas may have to be afterburned to reduce concentration of undesirable compounds in the offgases discharged to the atmosphere.

REFERENCES

1. Bień, J.B., and Bień, J.D., 1998, Unieszkodliwianie osadów ściekowych za pomocą metod termicznych" in: materiały konferencji Naukowo-Technicznej *Osady ściekowe w praktyce*. Częstochowa-Ustroń.
2. Kozłowska, B., 1998, Dlaczego piroliza, in: materiały konferencyjne *Termiczna utylizacja odpadów*, Poznań.
3. Barański, A., 1993, Piroliza - nowa technologia unieszkodliwiania odpadów. *Eko-Problemy 2.*
4. Sobolewski, A., and Haberski, A., 1995, Termiczne unieszkodliwianie zużytego sprzętu medycznego jednorazowego użytku, Sympozjum *Spalanie odpadów przemysłowych - technologie i problemy*, Łódź.
5. Ammann, P., 1995, Advantages of pyrolysis in recycling of dry batteries: The Recytec Process, *Congress Proceedings R'95 Recovery, Recycling, Reintegration*. Geneva.

6. Pinto, F., Gulyurtlu, I., Goncalves, M., and Carbita I., 1995, A new perspective for plastics recycling, thermolysis of plastic in an inert atmosphere", Congres Proceedings R'95 Recovery, Recycling, Reintegration. Geneva.
7. Tesch, M., and Moergeli, R., 1995, Termische Behandlung und Verglasung von Hydroxide und Lackchlammen; *Abfallwirtschafts Journal*, 5-7: 302-306.
8. Campbell, H.W, and Martinoli, D.A, 1990, Canada's Oil-From-Sludge Technology; Conference Proceeding *Status of Municipal Sludge Management for the 1990s.*
9. Bień, J.B., and Bień, J.D., 1996, Charakterystyka koksiku otrzymanego w procesie unieszkodliwiania osadu ściekowego w technologii OFS, Materiały Konferencji Naukowo-Technicznej *Wykorzystanie osadów ściekowych - techniczne i prawne uwarunkowania*. Częstochowa.
10. Bień, J.B., Bień, J.D., and Nowak, W., Próby wykorzystania technologii OFS do unieszkodliwiania osadów ściekowych, *Gaz, Woda i Technika Sanitarna 1/99.*

CATALYTIC LIMITATION OF ORGANIC CHLORIDE AND SULPHUR EMISSION IN THE THERMAL WASTE UTILISATION PROCESS

Marek Kaźmierczak
Institute of General and Ecological Chemistry, Technical University of Łódź, 90-924 Łódź, 36, Żwirki Str., Poland, e-mail: marekazm@ck-sg.p.lodz.pl

Key words: chlorine and sulphur organic wastes, waste treatment, catalytic oxidation

Abstract: The effect of temperature on thermocatalytic oxidation of 1,2-dichloropropane, diphenyl sulphone and thiazole was investigated. Platinum and copper-zinc catalysts were used in the experiments. The efficiency of the treatment process was estimated by measuring the concentrations of polychlorinated dibenzo-p-dioxins and dibenzofurans in the oxidation products of a chloroorganic compound, and the concentrations of sulphur dioxide in the oxidation products of an organic sulphur compound. For both oxidation processes, increasing the temperature to the range of 350 to 55°C was advantageous. The results of the investigations demonstrate the potential of thermocatalytic oxidation for treatment of waste mixtures containing organic compounds of both chlorine and sulphur.

1. INTRODUCTION

Minimisation and proper disposal of wastes are very important considerations in determining economic. Combustion is one of methods used for treatment of hazardous industrial wastes, including chloro-organic wastes. For these wastes the process is carried out at a sufficiently high temperature (1200 – 1500°C), at the contact time of greater than 2 seconds, and with quick cooling of combustion gases in order to limit the formation of very toxic polychlorinated dibenzo-p-dioxins (PCDDs) and polychlorinated dibenzofurans (PCDFs).

Combustion of wastes is applied in many countries, including Poland. For example, Chemical Plant "Rokita" SA in Brzeg Dolny, and Nitric Plant in Włocławek treat wastes containing chloro-organic compounds by combustion at temperatures of 1200°C or higher[1,2]. During this process, organic mass is reduced, chlorine is recovered in the form of hydrochloric acid, and water vapour is produced with application of heat from combustion gases. Exploitation of the system for non-catalytic combustion of wastes requires additional fuel and constructive materials resistant to corrosion at high temperatures. The additional expense involved for fuel and material limits the application of the process.

Thermocatalytic oxidation enables efficient treatment of hazardous organic wastes from an economic point of view. Application of the proper catalyst significantly decreases temperature in the reaction zone from 1200°C to about 400°C without environmental pollution from combustion products. A catalyst can be used for combustion of organic compounds after their preliminary treatment in a non-catalytic system.

Thermocatalytic oxidation for treating hazardous industrial wastes has been investigated by the Chemical Technology and Environmental Protection Group of the Institute of General and Ecological Chemistry at the Technical University of Łódź.

New efficient methods of treatment or destruction of various industrial wastes were developed as the result of these investigations. Some of these methods have been successfully applied or variations of these methods were supplied to industrial plants interested in them[3, 4].

Chlorine compounds occurring in liquid industrial wastes, such as 1,2-dichloropropane (DCP), were found to be interesting subjects for our investigation.

Thermocatalytic oxidation of organic chlorine compounds is a difficult task because hydrogen chloride and chlorine formed during the process can have an unprofitable effect on a catalyst. Moreover, they must be removed from combustion gases before their release to the atmosphere. The most effective strategy would be to prevent the formation of polychlorinated PCDDs and PCDFs.

Thermocatalytic oxidation of organic sulphur compounds occurring in liquid industrial wastes is also difficult. Sulphur oxide reaction products can contaminate the catalyst, and they also cannot be released to the atmosphere with combustion gases. Diphenyl sulphone (DPS) and thiazole were the sulphur-containing compounds most frequently used.

In experiments, different catalysts were used: platinum PA-2 containing 0.06% of Pt spread on the surface of Al_2O_3, platinum KP-910 containing 0.10% of Pt and copper-zinc TMC-2. The platinum catalyst PA-2 was produced by the Institute of Heavy Organic Synthesis "Blachownia" in Kędzierzyn Koźle, catalyst KP-910 - by Technical University of Wrocław, and copper-zinc catalysts - by Nitric Plants in Tarnów. Platinum catalysts were reduced to the grain size below a diameter of 0.5 mm. The grain-size distribution of oxide catalysts was 0.75 to 1.02 mm.

The results of investigations on thermocatalytic oxidation with application of the mentioned compounds and catalysts were presented in several papers[5-10]. Presented results allow a comparison of the thermocatalytic reaction process with the process carried out without catalyst.

The purpose of our investigations was to prove that application of a catalyst to the combustion processes of organic wastes is beneficial. The effect of temperature on the efficiency of thermocatalytic oxidation and on the formation of toxic PCDDs, PCDFs and sulphur dioxide emissions were investigated as a basic criterion for process evaluation.

2. EXPERIMENTAL

Experiments were carried out in quartz flow reactors with a diameter of 1.2 and 1.7 cm, both heated electrically. A vapour-gaseous mixture obtained in an evaporator heated to 160°C was pumped to the reactor containing a catalyst at a rate ensured the required contact time. A reaction mixture was formed as the result of the evaporation of 1,2-dichloropropane aqueous solution at concentration of 0.6 g/l (supplied at a mass flow rate of 32.5 g/h); DPS aqueous-ethanol or thiazole aqueous solution at concentration of 0.5 g/l (supplied with rate of 9 g/l) and air pumped at a volumetric flow rate of 15 to 120 l/h. Most of experiments were carried out in the temperature range from 200 to 700°C. Cooled combustion gases and aqueous condensate were the reaction products.

In the reaction products, concentrations of substrates, some intermediate and final products, were determined. The concentrations of DCP, DPS and thiazole were determined using a gas chromatograph equipped with flame ionisation detector. Total organic carbon in the oxidised solution and in condensate was analysed using TOC analyser produced by Shimadzu. The concentration of formaldehyde was determined by a colorimetric method with phenylhydrazine hydrochloride and the concentration of chlorides – was analysed using the Mohr or turbidimetric method. The concentration of carbon oxide and chlorine in combustion gases was analysed using an automatic gas analyser GA-20 produced by Madur.

3. RESULTS AND DISCUSSION

The results of experiments are presented in Figures 1-4. Figure 1 presents a dependence of efficiency of DCP oxidation vs. process temperature. Time contact was 0.36 s[11].

DCP was not present in combustion gases at temperatures higher than 375°C if the platinum catalyst was applied, and at temperatures higher 400°C in the presence of the copper-zinc catalyst. However, combustion gases contained intermediate products of oxidation. As shown in Figure 2, formaldehyde was present in combustion gases from the oxidation process with platinum catalyst PA-2 up to temperature of 450°C. Carbon monoxide was found in combustion gases from this process at temperatures lower than 500°C.

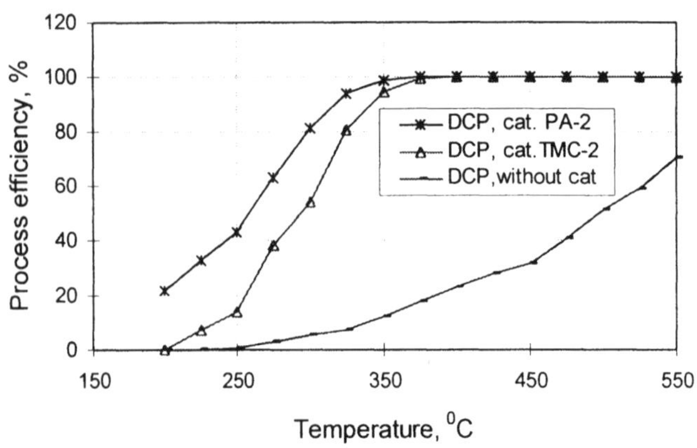

Figure 1. Efficiency of DCP oxidation versus process temperature.

The basic criterion for measuring process efficiency was the content of PCDDs and PCDFs in the combustion gases. Table 1 shows the effect of temperature on PCDD and PCDF emission concentrations[12]. The analyses were carried out at the Institute of Pulp and Paper in Łódź, in Doctor Małgorzata Michniewicz's laboratory. The concentrations of PCDD and PCDF decreased with the increase in the temperature of KP-910 catalyst from 375 to 550°C. Similar experiments were also carried out without application of a catalyst in the temperature of 800°C. In the thermocatalytic process, the concentration of dioxins was lower.

Figure 3 illustrates the realtionship between process temperature and the efficiency of DPS and thiazole oxidation on process temperature with the application of platinum and copper-zinc catalysts[13], and thiazole oxidation without a catalyst.

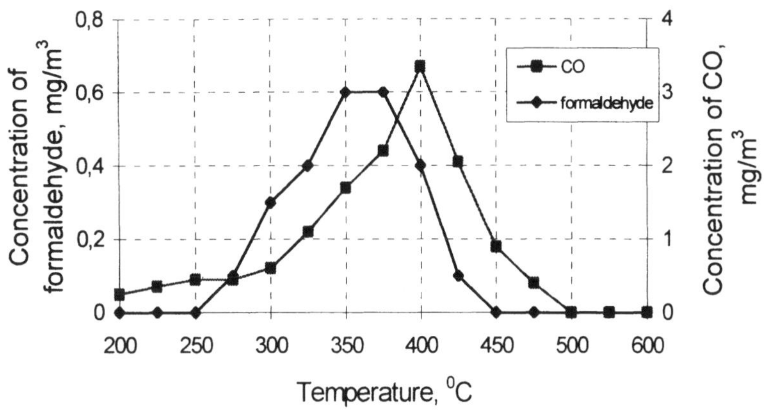

Figure 2. Concentrations of formaldehyde and carbon monoxide versus process temperature.

Table 1. Oxidation of 1,2-dichloropropare. Emission of PCDDs and PCDFs.

Type of process	Temperature (^0C)	TEQ (ng/m^3)
in presence of	375	0
platinum catalyst	450	0.0003
KP-910	550	0
without catalyst	800	0.051

The dependence of sulphur dioxide concentration on temperature and type of a catalyst is presented in Figure 4. The concentration of sulphur dioxide increases with the increase in temperature, and it reaches its maximum value in the temperature range 450 to 550°C. This figure shows also dependence of sulphur dioxide concentration vs. temperature in the non-catalysed process. Such the comparison allows to conclude that the catalyst causes an increase in the content of SO_2 in combustion gases. However, consideration of sulphur dioxide concentration as a function of conversion degree proves that this dependence is similar to the dependence obtained for the process with and without application of the catalyst[13].

The concentration of sulphur dioxide is not the only measure of combustion efficiency. Consideration of a sulphur balance allows one to conclude that if the combustion process is efficient then organic sulphur cannot occur in combustion gases. Thus, the sum of sulphur oxides concentration should be constant in combustion gases. If the combustion process is carried out in such a way to minimise concentrations of sulphur dioxide and absorption of acid compounds of combustion gases with sulphate production, then platinum or copper-zinc catalyst can be applied. If minimisation of the sulphur trioxide concentration in order to prevent the corrosion of apparatus is the main purpose, then another catalyst (e.g. one containing titanium dioxide as a main component) should be applied[14].

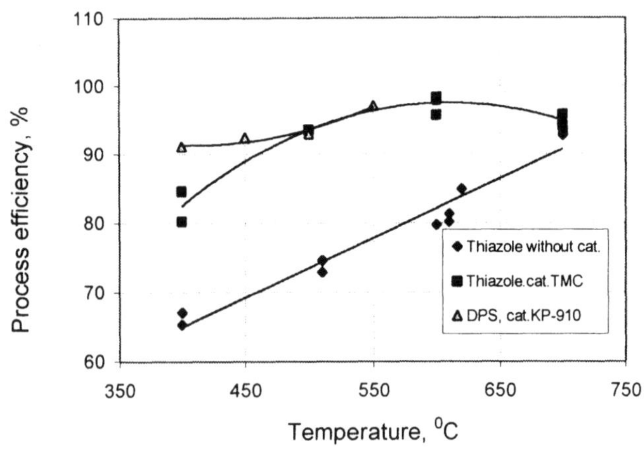

Figure 3. Efficiency of DPS and thiazole oxidation versus temperature and type of the catalyst.

The application of the copper-zinc catalyst and proper parameters of the process enables also to carry out the oxidation process simultaneously with the process of adsorption of sulphur oxides[13].

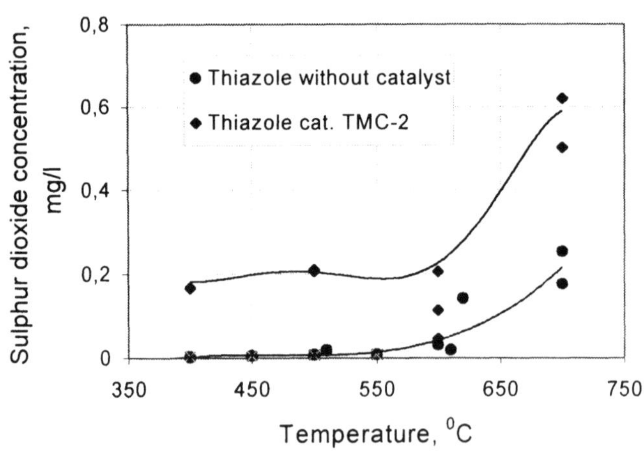

Figure 4. Sulphur dioxide concentration versus process temperature.

Minimisation of the content of organic chlorine compounds (high oxidation efficiency and low concentration of PCDDs and PCDFs) and organic sulphur compounds (high efficiency of oxidation process) requires the temperature of the process be maintained at about 550°C.

4. CONCLUSIONS

Results of experiments with 1,2-dichloropropane, DPS and thiazole with application of platinum (PA-2 and KP-910) and copper-zinc (TMC-2) catalysts demonstrate that these catalysts enable a significant decrease in the process temperature, from 1200-1500°C without a catalyst, to 550°C with a catalyst, without a decrease in its efficiency. At a short contact time of 0.01s, the efficiency of the oxidation process increased with an increase in temperature and reached the value of about 97% at 550°C. At this temperature, no intermediate products of DCP oxidation occur under experimental conditions, with a contact time of 0.36 s. At this temperature on the other hand, low concentrations of sulphur dioxide are detected in the oxidation process of organic sulphur compounds.

Considering simultaneous utilisation of organic chlorine and sulphur compounds, it is convenient that the effect of temperature on the process and optimum temperature is similar.

Finally, it should be that thermocatalytic oxidation must be followed by removal of hydrogen chloride and sulphur oxides from combustion gases.

ACKNOWLEDGEMENT

This work was supported by Polish State Committee for Scientific Research – grants No 0621/T09/97/13 and No 0152/T09/98/14.

REFERENCES

1. Doroczyński, A., Jóźwiak, T., 1997, Proceeding of III International Conference *Spalanie odpadów – technologie i problemy*. PŁ WIPiOŚ, Łódź, Poland.
2. Przondo, J., Rogala, J., 1996, *Przem. Chem.*: 98-101.
3. Gorzka, Z., Kaźmierczak, M., and Filipiak, E., 1984, Studies in Environmental Science. Vol 23: 437-443. Elsevier, Amsterdam.
4. Kaźmierczak, M., Gorzka, Z., and Michalska-Jednoralska, A., 1995, *Waste Management* *8*: 657-661.
5. Gorzka, Z., Kaźmierczak, M., Socha, A., and Michalska-Jednoralska, A., 1992, *Waste Management*: 345-348.

6. Żarczyński, A., Gorzka, Z., and Kaźmierczak, M., 1997, *Ochr. Powietrza i Probl. Odpadów* 1: 10-13.
7. Gorzka, Z., Żarczyński A., Kaźmierczak, M., and Miedzianowska, G., 1996, Proc. of IV Ogólnopol. Symp. *Ochrona Powietrza w Przemyśle*: 114-119, Łódź Poland.
8. Gorzka, Z., Kaźmierczak, M., and Żarczyński, A., 1998, Chemistry for Protection of the Environment 3, in: *Environmental Science Research* Vol. 55: 143-147. Plenum Publishing.
9. Kaźmierczak, M., Gorzka, Z., Żarczyński, A., and Miedzianowska, G., 1998, 13th Inter. Congr. Of Chem. And Proc. Eng. CHISA'98, Summaries 1: 82. Praha.
10. Kaźmierczak, M., Żarczyński, A., and Gorzka, Z., 1998, Proc. of III International Symp. Dioxyny-Człowiek-Środowisko: 1-6. Kraków Poland.
11. Kaźmierczak, M., Paryjczak, T., Gorzka, Z., and Żarczyński, A., 1999, European Meeting on Chemical Industry and Environment: 79-84. Kraków Poland.
12. Gorzka, Z., Kaźmierczak, M., Paryjczak, T., and Żarczyński, A., 1999, Thermocatalytic Oxidation of Chloro-Derivatives of Propane Occurring in Industrial Wastes. In: Proc. of XII International Conference *Chemistry for the Protection of the Environment*, suppl. In print. Nanjing China.
13. Kaźmierczak, M., 1999, Application of Catalystes in a Treatment Process of Organic Sulphur Compounds. In: Proc. of XII International Conference *Chemistry for the Protection of the Environment*, suppl. In press. Nanjing China.
14. Pająk, T., 1997, Proc. of III International Conference *Spalanie odpadów-technologie i problemy*: 277-286. WIPiOŚ. Łódź Poland.

EVALUATION OF GASEOUS EMISSION FROM ENERGY GENERATING FACILITIES

Ryszard Gierżatowicz
Institute of Environmental Protection Engineering, Lublin University of Technology, 40, Nadbystrzycka Street, 20-618 Lublin, Poland

Key words: gaseous emission, ambient air pollution, ambient pollution evaluation

Abstract: The most recent Polish regulations concerning the establishment of permissible values of pollutant emissions from energetic processes are simpler than those applied previously. Although in some cases the requirements to calculate the emission pattern around the source still exist. Paper presents review of the commonly used procedures for such calculations, together with sample calculations. It was found that considered procedures give the similar results, although there is a need to establish the only one recommended formula.

1. INTRODUCTION

The regulations recently brought into force in Poland concerning the establishment of permissible values of pollutant emissions from energetic processes[1,2, 3, 4] are simpler than those applied previously. For all categories of boiler, these designate permissible emissions as the concentrations of these pollutants in 1 m^3 of waste gases as converted to normal conditions, the situation for "dry" waste gases and with a concentration of oxygen in the gases defined in advance.

The legal regulations do not require calculations to be made of the breakdown of ambient pollution concentrations in the air surrounding energetic sources, even though these are needed in order to estimate whether permissible values for the concentrations brought about by permissible emissions are complied with. In the light of this, such calculations should be made in some cases at least, since the technical conditions for the evacuation of waste gases from existing or planned sources to the atmosphere are not always appropriate. This is especially true where there are tall residential buildings in the vicinity of a stack or chimney.

A further element not set out in the regulations concerns whether the checking of compliance conditions for the concentrations generated by permissible emissions from sources in general - and energetic sources in particular - is to proceed by way of measurement or also through calculation[5]. Though inexpensive, the calculated prognosis provides an insight into whether the bringing into operation of an emission source will need to be followed by measurement to check pollutant levels in the air, or whether this can safely be resigned from as unnecessary.

The calculation of the breakdown of pollutant concentrations in the vicinity of an energetic source requires prior determination of the permissible amounts of pollution expressed in mg/m^3 of dry waste gases (in agreed conditions and with an amount of oxygen set in advance) for emission of these pollutants expressed for example in g/s. This in turn requires familiarity with the volume (intensity of flow) of the waste gases. In drawing up documentation, it is necessary to take account of both the volume of waste gases in statutorily-defined conditions (to denote the emission as related to time), as well as the volume of wet waste gases, in order to define the speed of waste gases at the point of discharge from the flue. The precision with which permissible concentrations can be converted to permissible emission is dependent on the accuracy of the input data characterising the fuel, as well as on the mathematical apparatus used in making the calculations.

The further part of the article presents several of the calculation procedures applied most willingly in Poland, along with appropriate examples of their use. The aim will be to point out the differences arising in both the procedures it and the results.

The methods involving calculation always give results that are rather higher than the actual volume of waste gases since they fail to account for incomplete combustion. In the case of coal, the difference between the value for the real volume of waste gases and the calculated value may even reach 10%. Differences associated with incomplete combustion may only be determined when full heat measurements for a boiler are made. A further factor impacting upon the volume of waste gases is the content of water vapour in the air, which varies and may only be approximated.

2. THE PRESENTATION OF CALCULATION PROCEDURES

The basis for calculations is the general characterisation of fuels usually provided by their producers or distributors and including the calorific value, total or combustible sulphur content and the ash content. Recently, producers of coal have also included details of its chemical composition in their documentation.

The procedures usually applied in calculating the volume of waste gases initially base themselves on either the chemical composition of the fuel or its calorific value. Formulae relate to 1 kg of solid or liquid fuel, as well as to 1 Nm^3 of gaseous fuel. The amount of waste gas is given in Nm^3 (conventional cubic metres, otherwise known as normal cubic metres).

2.1 Calculation Proceeding from the Chemical Composition of Fuel

In the case of solid and liquid fuels, the theoretical demand for air in the combustion of 1 kg of fuel amounts to[6]:

$$L_t = 8.88 \cdot C + 26.44 \cdot H + 3.32 \cdot S - 3.33 \cdot O \qquad [Nm^3/kg]$$

where the symbols for the elements C, H, O and S denote their contributions by mass in the fuel [kg/kg].

The unit volume of dry waste gases (excluding nitrogen) is:

$$V_S = 1.85 \cdot C + (g - 0.21) \cdot L_t \qquad [Nm^3/kg]$$

The unitary volume of wet waste gases (i.e. dry waste gases increased by the value for the content of water vapour from the combustion of hydrogen, from moisture in the air and from moisture contained in the fuel) is:

$$C_{SW} = V_S + 11.11 \cdot H + 1.24 \cdot w + g \cdot L_t \cdot 1.6 \cdot x \; [Nm^3/kg]$$

where:

x is the content of moisture in the air [kg/kg]

w is the moisture content of the fuel [kg/kg]

g is a coefficient for the excess of air [dimensionless number]

The term $g \cdot Lt \cdot 1.6 \cdot x$, defining the amount of water vapour (moisture) introduced with the air may possibly be substituted by the expression $g \cdot Lt \cdot 1.24 \cdot y$, where:

y is the maximal content of water vapour in the air, which may be presented in the following manner as a function of temperature[7]:

Air temperature [°C]	Maximal content of water vapour [g/m^3]
-30	0.0005
-20	0.0011
-10	0.0024

Air temperature [°C]	Maximal content of water vapour [g/m^3]
0	0.0048
10	0.0093
20	0.0171
30	0.03

recalling the need to convert the volume Lt for the real temperature of the air.

It is easiest to determine the coefficient for the excess of air from the formula: $g = 21/(21 - O)$
where "O" denotes the percentage content of oxygen in waste gases as taken on the basis of the Regulation of the Minister of Environmental Protection, Natural Resources and Forestry of 8.09.1998[5].

The above formula is derived from that applied in the procedures presented below:

$G = V^t_{ss}/V^t_p \cdot (O/21 - O) + 1$

on the assumption that the expression V^t_{ss}/V^t_p is close to unity[8].

The value g may be read off from Fig. 137-28 on page 192 in the guide from Recknagel *et al.*[14].

In the case of gaseous fuels, the theoretical demand for oxygen amounts to:

$$L_t = 4.762(CO + H)/2 + 2CH_4 + 3C_2H_4 + 3.5C_2H_6 + + \Sigma (n + m/4) \cdot C_nH_m - O \qquad [Nm^3/Nm^3]$$

where the proportions of the different gaseous components are expressed in [$\mu m^3/\mu m^3$of gaseous fuel. The term $\Sigma(n + m/4) \cdot C_nH_m$ concerns hydrocarbons other than CH_4, C_2H_4 and C_2H_6 which may possibly be contained in the gaseous fuel.

The volume of wet waste gases (but excluding the moisture supplied from the air) is determined by the formula:

$$V_{SWl} = g \cdot L_T + 0,5(CO + H) + \Sigma (m/4) \cdot C_nH_m + CO_2 + O_2 + N_2 = \text{“the volume of gaseous fuel"} + g \cdot l_T - 0,5(CO + H) - \Sigma (1 - m/4) \cdot C_nH_m \qquad [Nm^3/Nm^3]$$

while the term $0.5 (CO+H) - \Sigma (1 + m/4) \cdot C_nH_m$, expressing the change in the volume of gas during combustion may be omitted. "The volume of gaseous fuel" is obviously 1 μm^3.

In turn, the volume of water vapour in waste gases that derives from the fuel amounts to:

$$V_{wl} = H + \Sigma (m/2) \cdot C_nH_m \qquad [Nm^3/Nm^3]$$

In consequence, the volume of waste gases is equal to the difference:

$$V_S = V_{swl} - V_{wl}$$

On the other hand, the calculation of the volume of wet waste gases taking moisture from the air into account requires the adding to the value

Vswl of the volume of water vapour from the air, i.e. the value of the expression g·Lt·1.6·x:

$V_{SW} = V_{SWl} + g \cdot L_t \cdot 1.6x$

2.2 Calculation Proceeding from the Calorific Value of Fuel

In this case there are three partially-similar methods of calculation.

2.2.1. Method Using the Index CO_{2max}

Calculations require a knowledge of the value CO_{2max} (i.e. the maximum percentage content of CO_2 in dry waste gases, with complete combustion and g equal to 1). The values of CO_{2max} for different fuels are[8]:

coke - 20.7%
hard coal - 18.5-19.3%
brown coal - 18.5-19.7%
fuel oil - 15.5-16%
diesel - 15.5%
peat - 19.2%
wood - 20.4-20.5% [9].

The value CO_{2max} for Polish hard coals is also related by a function to the content of volatile matter in coal:

Content of volatile matter [%]	20	25	30	35	40	45
CO_{2max} [%]	18.63	18.6	18.7	18.85	19.0	19.2

As boiler coals have contents of volatile matter in the range 30-45%, the usual values for CO_{2max} are in the range 18.7-19.2%. The above data may be adopted successfully for calculations, all the more so since characterisations of coal detail the content of volatile matter.

In wet waste gases, the value CO_{2max} may also be calculated from the following formula[10]:

$CO_{2max} = [1.87 \cdot C/(8.86 \cdot C + 21H - 2.6O)] \cdot 100$ [%]

The overall formula for the volume of dry waste gases is presented as follows:

$V_S = V^t_{ss} + (g - 1) \cdot V^t_p$ [Nm^3/kg]

where:

V^t_{ss} is the theoretical volume of dry waste gases with n=1 and complete combustion [in $\mu m^3/kg$]

V^t_p is the theoretical volume of air needed for combustion [$N\mu m^3/kg$]

g is the coefficient for the excess of air

Albeit with:

$V^t_{ss} = V^t_p \cdot (CO_{2max} + 67.5)88.5$ [Nm³/kg]

where in turn:

for solid fuels:

$V^t_p = 0.2412(Q^r_j - 4606)/1000 - (A_r - 6)/169$ [Nm³/kg]

for liquid fuels: $V^t_p = 0.2957(Q^r_j - 4606)/1000$ [Nm³/kg] [11]

where:

Q^r_j is the calorific value of the fuel in kJ/kg

A_r is the ash content of the fuel in [%]

To calculate the volume of wet waste gases it is necessary to add to value V_s the volumes of water vapour from the combustion of the fuel, from the moisture content of the fuel and from the air supplied for combustion, which is to say the value of the expression:

$V_{pw} = 11.11H + 1.24w + gV^t_p \cdot 1.6x$ [Nm³/kg]

where: x is the amount of moisture in the air in kg/kg

i.e.:

$V_{sw} = V_s + V_{pw}$

The coefficient for the excess of air may be determined most readily from the formula:

$G = V^t_{ss}/V^t_p \cdot (O_2/21 - O_2) + 1$

2.2.2. Method Using Formulae after Cholewa and Kwiatkowski[12]

The amount of waste gases is calculated according to the following formula:

$V_{sg} = V^t_{sp} + (g - 1) \cdot V^t_p$ [Nm³/kg or Nm³/Nm³ soil]

For solid fuels, the theoretical volume of the waste gases arising from the combustion of 1 kg of fuel in normal conditions amounts to:

$V^t_{sp} = 0.95W_u/1000 + 1.375$ [Nm³/kg]

while the demand for air in the combustion of 1 kg of solid fuel [Nm³/kg] is obtained from:

$V^t_p = 1.012W_u/(1000 + 0,5)$ [Nm³/kg]

where:

W_u - is the calorific value of the fuel in [kcal/kg]

g is the coefficient for the excess of air:

$g = V^t_{ss}/V^t_p[O_2/(21 - O_2)] + 1$

Correspondingly, for liquid fuels, we have:

$V^t_{sp} = 1.11W_u/1000$ [Nm³/kg]

$V^t_p = 0.88W_u/(1000 + 1.7)$ [Nm³/kg]

and for gaseous fuels:

$V^t_{sp} = 1.09W_u/(1000 + 0.446)$ $[Nm^3/Nm^3]$
$V^t_p = 1.09W_u/(1000 - 0.28)$ $[Nm^3/Nm^3]$

It is not certain whether the calculated volumes of waste gases pertain to the dry or wet states, though it is assumed to be the latter, albeit excluding moisture from the air. Making this assumption, it is still therefore necessary to deduct the value for the expression $H + \Sigma(m/2)\cdot C_nH_m$ - in order to obtain the volume of dry waste gases, as well as to add the value from the expression $g\cdot V^t_p\cdot 1.6x$ - to calculate the volume of wet waste gases. Thus for gaseous fuels we obtain:

$V_s = V_{sg} - H - \Sigma(m/2)\cdot C_nH_m$
$V_{sw} = V_{sg} + gV^t_p \cdot 1.6x$

While for solid and liquid fuels we have:

$V_s = V_{sg} - 11.11H - 1.24w$
$V_{sw} = V_{sg} + gV^t_p \cdot 1.6x$

2.2.3. Method Using Formulae after Recknagel[13]

The calculation procedure resembles that given in 2.2.2., though the set of approximated formulae (after Rosin and Fehling) for calculating the volume of wet waste gases is rather different:

$V_{sw1} = V_{Amin} + (g - 1)\ L_t$

wherein for solid fuels:

$L_t = 0.241H/(1000 + 0,5)$ $[Nm^3/kg]$

with H being the calorific value in kJ/kg

- for oils:
  - $L_t = 0.203H/(1000 + 2,0)$ $[Nm^3/kg]$
  - $V_{Amin} = 0.265H/1000$ $[Nm^3/kg]$
  - where H is the calorific value in kJ/kg
- for gases of low calorific value like blast-furnace gas, producer gas and water gas ($H<12{,}500\ kJ/\mu m^3$)
  - $L_t = 0.209H/1000$ $[Nm^3/kg]$
  - $V_{Amin} = 0.173H/(1000 + 1.0)$ $[Nm^3/kg]$
  - where: H is the calorific value in kJ/Nm^3
- for gases of high calorific value, e.g. town gas, coke-oven gas, natural gas and oil gas ($H>12{,}500\ kJ/Nm^3$)
  - $L_t = 0.260H/(1000 - 0.25)$ $[\mu m^3/\mu m^3]$
  - $V_{Amin} = 0.272H/(1000 + 0.25)$ $[\mu m^3/\mu m^3]$
  - where: H is the calorific value in kJ/Nm^3

Since the above formulae serve in calculating the volume of wet waste gases (in a presumption including water vapour from fuel moisture content and the combustion of hydrogen), the volume of water vapour should be deducted from those obtained using them, such that the value of the expression is:

a) for solid and liquid fuels:
$11.11H + 1.24w$

b) for gaseous fuels:
$H + \Sigma(m/2)C_nH_m$

In turn, in taking account of air humidity, it is necessary to add to the obtained Vsw1 values, a value from the expression g·Lt·1.6x, thereby giving, in the case of solid and liquid fuels:

$V_s = V_{sw1} - 11.11H - 1.24x$

$V_{sw} = V_{sw1} + g{\cdot}L_t \cdot 1.6x$

and in the case of gaseous fuels:

$V_s = V_{sw1} - H - \Sigma(m/2)C_nH_m$

$V_{sw} = V_{sw1} + g{\cdot}L_t \cdot 1.6x$

3. SAMPLE CALCULATIONS

Calculations were made for fuels of the following characteristics:

a) solid fuel.

This was hard coal in the form of pea coal from the WESOŁA mine in Mysłowice, which was characterised by the mine as follows:

Calorific value	= 26000-27000 [kJ/kg]	= 26500 [kJ/kg] average
Ash content	= 6.0 – 10.0 [%]	= 0.08 [kg/kg solid] average
Total sulphur content	= 0.41 – 0.6 [%]	= 0.005 [kg/kg solid] average
Moisture content	= 7.0 – 9.0 [%]	= 0.08 [kg/kg solid] average
Content of volatile matter	= 30 – 31 [%]	= 0.305 [kg/kg solid] average
C content	= 71.39 [%]	= 0.7139 [kg/kg solid]
H content	= 4.19 [%]	= 0.0419 [kg/kg solid]
N content	= 1.11 [%]	= 0.0111 [kg/kg solid]
O content	= 10.15 [%]	= 0.1015 [kg/kg solid]

The assumed water content of the air was 0.005 kg/kg
The required content of oxygen in waste gases was 6%
The adopted value for CO_{2max} was 18.75%

b) liquid fuel

The example fuel here is EKOTERM fuel oil. Adopted in the face of the lack of a precisely-defined chemical composition were the standard C =

0.85 kg/kg and H = 0.15 kg/kg, a calorific value of 41,500 kJ/kg, a density of 0.9 g/cm³, a sulphur content of 0.003 kg/kg and an ash content of 0.0001 kg/kg. The required content of oxygen in waste gases is 3%. The water content of the air was assumed to be at 0.005 kg/kg. The value of CO_{2max} adopted was in turn 16%.

c) gaseous fuel

The fuel chosen was natural gas of the following characteristics:

Calorific value	8643 [kcal/μm^3]	
Density	0.7475 [kg/μm^3]	
Total sulphur content	100 [mg/μm^3]	
Content of CH_4	96.56 [% obj.]	0.9656 [μm^3/μm^3]
Content of C_2H_6	1.43 [% obj.]	0.0143 [μm^3/μm^3]
Content of C_3H_8	0.42 [% obj.]	0.0042 [μm^3/μm^3]
Content of C_4H_{10}	0.1 [% obj.]	0.001 [μm^3/μm^3]
Content of C_5H_{12}	0.07 [% obj.]	0.0007 [μm^3/μm^3]
Content of CO_2	0.32 [% obj.]	0.0032 [μm^3/μm^3]
Content of N_2	1.1 [% obj.]	0.011 [μm^3/μm^3]
Content of H_2	0 [% obj.]	0 [μm^3/μm^3]
Content of CO	0 [% obj.]	0 [μm^3/μm^3]
Content of O_2	0 [% obj.]	0 [μm^3/μm^3]

The adopted value for air humidity was 0.005 kg/kg of air.
The required oxygen content in waste gases is 3%.

4. RESULTS OF CALCULATIONS AND CONCLUSIONS

The obtained results of the calculations are compiled in the table below:

Fuel	Solid		Liquid		Gaseous	
	V_s	V_{sw}	V_s	V_{sw}	V_s	V_{sw}
According to 2.1	9.8	10.45	12.60	14.37	10.21	12.29
According to 2.2.1.	9.46	10.10	12.01	13.78	-	-
According to 2.2.2.	9.78	10.42	11.17	12.93	9.51	11.60
According to 2.2.3.	9.61	10.25	11.16	12.93	9.77	11.86
Mean	9.66	10.31	11.92	13.50	9.83	11.92
Interval of occurrence of results in [%] of mean value:	-0,51 - +1.45	-2.04 - +1.36	-6.3 – + 5.7	-4.2 - +6.4	-3.24 - +3.86	-2.7 - +3.1
Difference between mean values V_{sw} and V_{sw} [%] of V_s value:	6.65		13.24		21.23	

There are differences between the results from particular calculation procedures for the same fuel (largest in the case of liquid fuel), but overall these are of limited significance from the point of view of typical engineering applications. It is not possible to discern any decidely superior procedures, although the results obtained on the basis of the chemical composition of the fuel are clearly higher than those gained using the other sets of formulae.

To convert the permissible amounts of pollutants expressed in mg/m^3 of dry waste gases (in normal conditions and with the given amount of oxygen assumed in advance) into the emission of these pollutants expressed for example in g/s, use is made of the calculated value Vs.

To calculate the outlet velocity of waste gases from a flue, use is made of the calculated value V_{sw}, obviously with a correction introduced for the real temperature conditions at the outlet.

It would seem that there is no danger of making a more major error in the calculation of value V_{sw} by increasing the calculated value V_s by the percentage given in the above compilation, and hence in relation to fuel type by 7, 13 or 21% (where the rounding-off accepted here simplifies the calculation slightly).

However, in general terms it would represent a favourable move if the Ministry of Environmental Protection, Natural Resources and Forestry were to introduce an obligation that one particular calculation procedure be applied.

REFERENCES

1. Regulation of the Minister of Environmental Protection, Natural Resources and Forestry of 3.09.1998 on the methods of calculating the situation as regards air pollution for existing and planned sources (*Dz.U.* 122/98, item 805).
2. Regulation of the Minster of Environmental Protection, Natural Resources and Forestry of 18.09.1998 on detailed principles by which to establish the types and amounts of polluting substances whose introduction into the air is permissible, as well as the requirements to be met by the documentation necessary for the issuing of a decision establishing the types and amounts of polluting substances permitted to be introduced into the air (*Dz.U.* 124/98, item 819).
3. Regulation of the Minister of Environmental Protection, Natural Resources and Forestry of 8.09.1998 on the introduction to the air of polluting substances from technological processes and technical operations (*Dz.U.* 121/98, item 793).

4. Regulation of the Minister of Environmental Protection, Natural Resources and Forestry of 28.04.1998 on permissible concentrations of polluting substances in the air (*Dz. U.* 55/98, item 355).
5. *Dziennik Ustaw* no. 129/98, item 819, paragraph 3.1
6. Recknagel, Sprenge, Houmann, and Schramek, 1994, Ogrzewanie i Klimatyzacja, EWFE, Gdańsk; Mizielińska, K., and Rubik, M., 1995, Ciepłownictwo, Poradnik, Źródła ciepła, Unia Ciepłownictwa
7. Król, Cz., 1981, *Hydrologia*, PWRiL Warszawa.
8. Mieszkowski, M., (editor), 1985, Heat and energetic measurements, WNT, Warsaw.
9. Kapitaniak, A., and Sztraube, J., Poradnik palacza, WNT, Warszawa.
10. Recknagel, Sprenger, Houmann, and Schramek, 1994, Ogrzewanie i klimatyzacja, EWFE, Gadańsk: 192.
11. Boise, W., 1957, Vom Brennstoff zum Rauchgas. Leipzing. Teubner Verlagsgesellschaft
12. Kwiatkowski, J., and Cholewa, L., 1980, Centralne ogrzewanie. Pomoce projektanta, Arkady, Warszawa.
13. Recknagel, Sprenger, Houmann, and Schramek, 1994, Ogrzewanie i klimatyzacja, EWFE, Gdańsk.

PLASMA TECHNOLOGIES FOR ENVIRONMENTAL PROTECTION

Krystyna Cedzyńska and Zbigniew Kolaciński
Technical University of Lodz, 18/22 Stefanowskiego Street, 90-924 Lodz, Poland

Key words: thermal plasma, waste destruction

Abstract: Efficient approach to the treatment of toxic and hazardous wastes is offered by thermal plasma. Because the arc plasma provides the means to operate the processes at much higher temperatures that are commonly achieved with conventional combustion it can be used to the treatment of the most stable chemical compounds. The target applications of thermal plasma technology are: treatment of the most toxic and hazardous wastes, clean fuel generation from waste and disposal of leftovers coming from conventional incineration plants and nuclear power stations. Some of the research results achieved in the Plasma Technologies & Environmental Protection Group at the Technical University of Lodz are here presented and discussed.

1. INTRODUCTION

Storage or landfilling of toxic and hazardous wastes causes a long-term problem of increasing dimensions. Among the treatment methods, the thermal treatment has gained the widest acceptance because it has proven to have the best destruction efficiencies of organic waste fraction. The present day incineration with the oil or gas burners is a well-developed technology. The process temperature is reaching 1500°K and the sophisticated off-gas cleaning system is effective enough to meet the gas emission standards. However the incineration has the following shortcomings:

- an inert control of the combustion process and long start-up and shutdown times can cause incomplete oxidation resulting in temporary hazardous gas products,

- the waste processing is a decreasing of its volume and mass only,
- the ash residue being a secondary waste is sometimes more toxic than the primary solid feed,
- the excess of air required for combustion results in large volume of greenhouse gases released to the atmosphere,
- dioxins, furans and heavy metals can be released as air borne pollutants in the ash, which needs to be landfilled with special care due to a danger of ground water contamination.

The above problems can be resolved by using plasma technology for thermal treatment of toxic and hazardous wastes.

2. THERMAL PLASMA WASTE TREATMENT POTENTIAL

2.1 Advantages

The thermal plasma is a source of high energy density with temperature of a few thousand degrees and sufficient ultraviolet radiation. This results in:

- fast reaction rates,
- high throughput in smaller reactors,
- heat generation independent of the chemical composition of the treated waste,
- avoidance of dioxins and furans formation due to oxygen free atmosphere,
- freedom to select optimum chemistry for the destruction process,
- melting of high temperature inorganic wastes,
- easy control (electricity is the energy source),
- rapid start-up and shut down,
- flexible treatment process.

2.2 Development and Treatment Strategies

Majority of the plasma equipment is well-tested in conventional applications such as: plasma cutting, welding, melting and spraying. The plasma furnaces are successfully used in metallurgical extraction and metal refining. The emerging technologies are plasma synthesis or plasma decomposition of materials and plasma waste treatment. The typical waste treatment reactors operate with DC plasma torches and DC or AC graphite-electrode systems. In some plasma reactors and for some wastes it is convenient to have the arc to be transferred to the material to be heated. The plasma torch operates then as one electrode, while the material to be heated is connected to the other electrode. In this arrangement the heat transfer is enhanced by electrode spot effect. Power rating of a single transferred plasma torch presently used is in average of from 1kW up to 20 MW. The maximum operating voltage is imposed by practical problems of stray arcing within the furnace, while the current limit is defined by electrode lifetime considerations. The state of the art torches are very durable, and in some cases an uninterrupted operation between servicing for exchange of electrodes can reach 1000 hours assuming inspection and adjustment is performed every 100 hours. The thermal efficiency can reach 73% for non-transferred arc torch and 90% for a torch with transferred arc. The key advantage of the non-transferred plasma torch is that the source of heat generated is electrically and chemically inert. The temperature of the plasma jet depends mainly on the working gas dissociation and ionization potential and slightly on the cooling conditions (arc rotation or turbulence in the furnace) and the gasified and melted waste medium. In the case of graphite arc, the power rating can be higher but the continuous electrode consumption at the rate of 1.5 kg per 1000 kg of waste must be considered as well as the fact that the plasma arc is the waste medium depended.

The general system of waste treatment with plasma can be considered as a combination of two plasma heaters. The primary unit is to treat the mixed solid/liquid waste i.e. polychlorinated biphenyl's (PCB) oil contaminated soil, containers with toxic substances, capacitors, etc. The process can be arranged s gasification and melting of inorganic (metal and ash) fraction of the solid/liquid waste. The secondary unit is the reaction chamber and it works under the "3T principle" which means that the three parameters: temperature, time and turbulence are optimized to evaporate the droplets and to break the molecules chains. This unit is responsible for final treatment of the gaseous products leaving the system. It employs the electric arc or plasma torch (if pyrolytic production of the fuel gas is assumed). A conventional burner can also be applied if combustion process is concerned.

In the case of processing the liquid waste only, the feed can be injected directly into the plasma jet however in the most cases a dilution of the oily fluid with some solvents is necessary to rich the best atomizing efficiency. To avoid formation of dioxins, the flue gas has to be cooled and quenched rapidly with the rate above 10^5 ^{0}K/sec and washed in an alkali scrubber for removing acids.

The plasma process on an industrial scale is a quite sophisticated and know-how packed technological system. The target applications of the thermal plasma systems presently are:

- treatment of the most toxic and hazardous industrial wastes,
- clean fuel generation from organic hazardous wastes,
- vitrification of solid leftovers from conventional incineration plants and nuclear power stations,
- vitrification of contaminated soil.

- The applications specified above will be demonstrated below following the research work conducted at the Plasma Technologies & Environmental Protection Group (PTEP) of the Lodz Technical University.

3. TOXIC AND HAZARDOUS WASTE

A major advantage of plasma processing is that the heat to the process can be provided in an atmosphere of any desired composition and reactivity. In practice there are only a few variations of chemical reactions available for thermal processing i.e.: pyrolysis, oxidation, reactions with hydrogen and water. To optimize the conditions of the reaction we have calculated the decomposition rate of chloroorganic compounds within the

range of temperatures from 500 to 5000^0K using the computer program based on minimization of the Gibb's free energy[1]. Those calculations included the systems of pure compounds or their mixtures with water, hydrogen and oxygen[2].

The plasma decomposition can be easily demonstrated on a well-known electrical insulation fluid - dichlorinated PCB, which belongs to the most toxic substances. Burning of PCB's causes serious problems concerning easy formation of dioxins and furans. Pyrolysis of PCB up to the temperature 1800^0K results in appearance mainly mixture of CH_4, HCl and H_2 (Fig. 1a). For temperature higher than 3200^0K, carbon in a form of soot is formed. This process is not desirable because the generated soot covers the electrodes and electrical insulation material. To eliminate this detrimental phenomenon H_2O in the form of steam has to be added (Fig. 1b). Water generates OH radicals, which provide an effective destruction mechanism, and also hydrogen atoms, which remove halogens. In the temperature about 1500^0K the clean (without soot) fuel gas composed of mainly CH_4, CO and H_2 is formed. The disadvantage of this process is the high-energy consumption, which however, can be recovered with the fuel gas calorific value. The PCB process temperature higher than 2000^0K (Fig. 1b) results in chlorine formation. It means that from the point of view of receiving the most clean flue gas, the PCB treatment process should be carried out at the temperature not exceeding 2000^0K. Another characteristic of any thermal destruction process is that many of the products are chemically reactive so as their temperature falls, they are free to recombine and to form higher molecular weight molecules. Therefore the rapid quenching of the flue gas is required (mostly in alkali scrubber for HCl removal). Our experiments with hazardous fluid and gas waste treatment including PCB's were done in a chemical reactor based on the rotating arc principle (ROTARC) and they had been reported elsewhere[3]. The destruction efficiency better than 99.99% had been easily achieved.

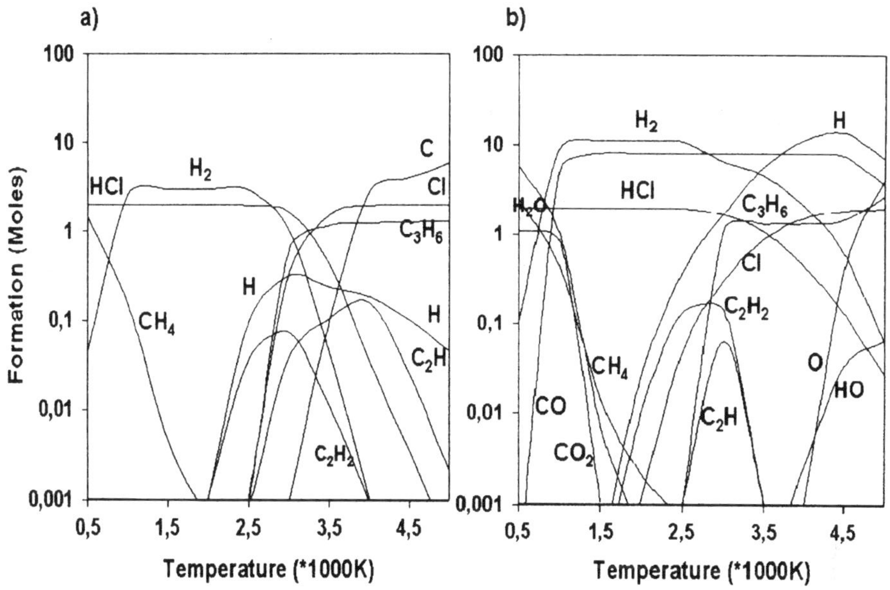

Figure 1. Dichlorinated PCB thermal decomposition: a) pyrolysis, b) reaction with water.

4. FUEL GENERATION

Thermal plasma is ideally suited for the massles heating of solid and liquid hydrocarbons to convert them into by-products having recoverable values. The gas mixture contains primarily hydrogen and carbon oxide, approximating reformed natural gas, but the molten slag-metal mixture solidifies into a non-leachable solid. The products rich in hydrogen and carbon oxide may be used in the manufacture of methanol, or used as fuel to generate steam for electric power or directly run the gas motor or turbine to generate electricity.

Gas fuel generation from wastes is very attractive and environmentally friendly process. Several computer analysis of the solid waste conversion into gas fuel was done in the PTEP Group[4]. As an example we present here some computation results for the plasma treatment of the hospital (patients room) waste (yellow bag) including of plastics, cellulose and non-organic metal and glass parts. To check the efficiency of gas fuel generation from this waste we have simulated the pyrolysis process – reported elsewhere[4] with limited oxidation carried out within temperature range 1000 – 2500^{0}C.

Figure 2. PTEP Group bench scale system for hospital waste treatment with plasma

We have assumed that in 1kg of dry waste are (by weight) 60% of plastics, 15% of silicon glass, and 10% of calcium glass, 10% of papers and 5% of metals. Moisture content of 15% was assumed in the calculations. In the case of pyrolysis in temperature below 1500^0C the main component of solid fraction was soot. Above 2000^0C the whole amount of oxygen has been used to convert the carbon into CO however, still more than 45% of soot was present in the system. It can be seen in Table 1 and Table 2, that the oxidation rate resulting in CO_{max} production may be achieved with relatively small amount of oxygen supplied (approximately 0.6 kg) mantaining the same calorific value of the product gas. The best temperature range for the plasma processing is from 1500 to 2000^0C, because in this range the glass fraction melts and encapsulates even refractory metals. In such process conditions the energy rate required for a plasma torch (including losses) is 1.2 kWh/kg, but about the same value of energy is in the fuel gas produced, which can be converted (however with some losses) into heat or electrical energy. There are two main advantages of fuel gas production in the plasma process compared to incineration in the oxygen starvation conditions: the highest possible calorific value, and vitrification of ash and slag due to high process temperature.

The PTEP has also built laboratory installations for the destruction of hospital wastes (Fig. 2). The laboratory studies confirmed high efficiency of the destruction of those wastes (99.99%). Computer simulations of technological system for a medium size hospital were made. The expected capacity of the installation according to the proposed solution is about 40 kg of wastes per hour. The electric energy requirement for the decomposition of hydrocarbons and dioxins as well as for the vitrification of ashes and heavy metals can be recovered in the produced fuel gases so that the energy balance of the system is positive.

Table 1. Products of plasma limited oxidation

Temperature	Oxygen at CO_{max}	CO	H_2	H_2O	CO_2	C
°C	kg	Mass percentage				
		Volume percentage				
1000	0.56	76.0	4.0	4.9	3.9	11.0
		72.767	22.991	3.196	1.046	
15000	0.69	84.5	7.7	3.8	2.8	1.0
		63.548	33.858	3.004	0.614	
2000	0.60	86.8	8.0	2.0	0.2	1.0
		63.443	35.100	0.974	0.044	

Table 2. Energy balance at plasma limited oxidation.

Temperature	Oxygen at CO_{max}	Net energy consumption	Calorific value	Energy in gas fuel
°C	kg	KWh	MJ/m^3	MJ kWh
1000	0.56	- 0.50	12.662	2.363 0.656
1500	0.69	- 0.35	12.672	3.063 0.851
2000	0.60	+ 0.90	12.673	3.188 0.886

5. VITRIFICATION

The incineration of certain wastes (industrial, medical, military) results in the formation of relatively highly toxic residues. The toxic leftovers (ash, slag, filter deposits, sedimentation residues) could be easily disposed in landfills assuming that they were first immobilized and converted into a non-leaching product. When the leftovers are heated to a sufficient temperature, their elements, including minerals and toxic heavy metals, melt and glassify. Even partial solidification (vitrification) of those residues requires the temperature above 1700^{0}K which is not available in most incinerators but easily reachable in thermal plasma reactors[5]. Temperatures of the order of 10000^0K are typical for arc in plasma furnaces and all inorganic residues can be solidified. The plasma ash vitrification system produces a chemically stable and mechanically resistant product. After vitrification the mineral product looks like a glassy, basalt structured lava (even of higher than basalt mechanical strength), and its main components are silicon, aluminium and calcium oxides in the form of chemically inactive compounds, resistant to leaching.

The DC plasma furnace was designed in the Technical University of Lodz and applied to treat the ash samples. The furnace was water-cooled and the arc-plasma system had a maximum output power of 150 kW. Samples of vitrified material in the form of lumps were tested using the Vicker's method to measure their hardness. The heavy metals leaching tests were performed using demineralised water with conductivity below $0.1 \cdot 10^{-9}$ S/cm. Extraction was done in shaker for 24 hours at room temperature. Concentrations of ions in the extracts were measured using the anodic stripping voltamperometric technique. Determination of oxides concentration in ashes was done using the method of X-ray fluorescence spectrometry.

The typical bottom ash from one of the Polish hospital incinerators was taken as representative sample of medical toxic residues. Since the ash was kept in an open air it contained about 10% humidity. As the filler we have used a fly ash produced by typical coal fired power plant of Lodz district. The morphology of the both ashes is presented in Table 3.

Table 3. Chemical analysis of the hospital incinerator bottom ash and the power plant fly ash.

Oxide	Content	
	Bottom ash [weight %]	Fly ash [weight %]
SiO_2	47.35	50.02
Al_2O_3	3.05	23.20
Fe_2O_3	7.35	9.25
Mn_3O_4	*	0.15
TiO_2	*	0.96
CaO	16.25	4.08
Na_2O	*	0.96
MgO	2.45	2.48
K_2O	*	3.40
P_2O	0.30	0.32
SO_3	0.48	0.52
Organic compounds	*	4.25

*) contents not measured

During each run typically 5 kg of mineral feed (homogenised ash) was melted and continuously overflowed. Casting in the water-cooled container gave the lumps of the final vitrified product. The solid material obtained after cooling was then analysed. Both the feed and the product weights were recorded for mass balance purposes.

The main compound of the both streams (bottom and fly ashes) was silica oxide. This influenced the melting point of the mixture being between 1550 – 1600^{0}C. The testing was done for three different ratios of bottom to fly ash (0.25, 0.50, 0.75). In all cases the cooled slag was homogenous and vitreous in nature (Fig. 3). The final product looks like a having basalt structure. The vitrified material mass was only 85% of mass of the material fed into the furnace for the mixture ratio 0.75 and 95% for the mixture ratio 0.25. The example of plasma vitrified product for 50/50% mixture of bottom ash coming from medical waste incineration and fly ash coming from the power plant is shown in Figure 3. The process of plasma vitrification of ash and slag ends in a chemically stable (Table 3) and mechanically resistant product[8], which is safe to store or it can be used as construction material in the form of bricks or roof tiles.

Figure 3. Mixture of bottom and fly ashes in comparison with plasma vitrified product.

Table 4. Analysis of heavy metals concentration in aqueous extract obtained from leaching test for bottom ash (medical incinerator), fly ash (power plant) and vitrified slag for mixture of bottom and fly ashes 50/50%.

Metal ion	Concentration in aqueous extract [mg/dm^3]		
	Bottom ash	Flay ash	Vitrified slag
Pb^{2+}	0.80	0.50	<0.005 *
Cu^{2+}	0.63	0.55	0.010
Ni^{2+}	0.84	0.28	<0.005 *
Zn^{2+}	0.52	0.66	<0.005 *
Cr^{3+}	0.05	2.26	<0.005 *
Mn^{2+}	0.42	0.78	<0.005 *

- below detectable limit
-

The plasma vitrification process can also be used for asbestos waste which is a hazardous, fibrous, dusty material posing danger to human beings such as lung cancer – mesothelioma. Nowadays-huge volumes of asbestos-based waste thermal insulation material wait for some kind of treatment to be converted into non-hazardous waste or a construction material. Thermal plasma can be easily employed to convert asbestos into a dense solid structure, which can be environmentally safe.

6. SOIL DECONTAMINATION FROM TOXIC COMPOUNDS

In Poland as well as in many eastern european countries a number of used capacitors and transformers containing mineral oil with polychlorinated biphenyls (PCB's) are stored. The practically unavoidable frequent leaks of oil with PCB's from transformers during their operation have contaminated the soil in numerous places. Most of the PCB's that were released to the environment adsorbed in soils and sediments where they still persist. In the case of a localized accident with PCB leakage the required response is to stop, as quickly as possible, the fluid penetration into the ground. However, for contaminated soil there are two preferred approaches:

- Vitrification "in situ" to convert the contaminated soil into non-leachable solid material, which is inert to the environment,
- Collection and transportation of the contaminated soil to a furnace for vitrification or to a chemical washing machine, from which extracted PCB's are transferred to the thermal treatment furnace ("ex situ" action).

In the PTEP Group several experiments were performed with thermal treatment of PCB polluted soil[9]. A 3-phase electric smelter working on principle of soil electrical resistance heating was employed for soil melting and gasifying of the organic fraction. The ROTARC plasma reactor has been successfully used to treat the off-gas coming from the electric smelter. In the case of washing the soil with a solvent, the liquid extract was directly injected into the plasma reactor for decomposition. The destruction rate was always better than 99.99%.

7. CONCLUSIONS

The main advantages of thermal plasma waste treatment are:

- complete pyrolysis of organic and hazardous wastes – no need of oxygen,
- vitrification of inorganic toxic leftovers – high temperature,
- high calorific value of gas fuel generated,
- smaller size of installation - high energy density
- minimum gas throughput - no excess gas for combustion required,

- rapid start-up and shutdown - easy automatic control.
- World-wide, there are numerous plasma system designs for treatment of all types of wastes[10]. Economical considerations limit their commercial applications to the most profitable actions. Presently they commercially operate: in Switzerland and Germany for low level nuclear waste vitrification, in France and the USA for asbestos waste vitrification, in the USA and Australia for hazardous waste treatment, in Japan and France for municipal fly ash vitrification. The most of installations is working in Japan because there is 70% of municipal waste incinerated and the ash can not be used as landfill. The latest available EEC statistics (Environmental Statistics Yearbook Eurostat 1996) show, that approximately 10 million tonnes of hazardous waste was produced within the Community. EEC regulations banning the disposal to landfill of toxic and hazardous wastes are expected to become effective in year 2005. Presently it is difficult to find a method of hazardous residues processing able to compete with plasma vitrification. This is the economically efficient way of returning this toxic waste to the environment in the form of friendly and useful construction material.

ACKNOWLEDGEMENT

The State Committee of Science (KBN) Grant No 8 T10A 031 16 sponsored this work.

REFERENCES

1. Eriksson, G., and Hack, K., 1990, ChemSage – A Computer Program for the Calculation of Complex Chemical Equilibria", Met. Trans. B. vol. 21B: 1013-1023.
2. Kolacinski, Z., and Cedzynska, K., 1997, Chlorinated Waste Destruction by Plasma Pyrolysis, Progress in Plasma Processing of Materials: 493-500. New York, Wallingford.
3. Cedzynska, K., Kolacinski, Z., Dokimuk, J., and Gaworczyk J., 1997, Waste Destruction in the ROTARC Reactor, Proc. VIII Int. Symp. SAP&ETEP: 326-330. Łodz.
4. Cedzynska, K., and Kolacinski, Z., 1998, Plasma Pyrolysis – a Novel Method of Waste Destruction, *Polish Journal of Applied Chemistry*, No. XLII, vol. 2: 143-149.
5. Kolacinski, Z., at al, 1991, Temperature Measurements of a Plasma Jet for Pyrolysis of Liquid Waste", Proc. X Int. Symp. Plasma Chemistry and Plasma Processing, Bochum :12-33.
6. Cedzynska, K., Kolacinski, Z., Izydorczyk, M., and Sroczynski, W., 1999, Thermal Plasma Treatment of Waste Incinerator Ash, *Progress in Plasma Processing of Materials*. Begell House Inc.: 707-711. New York.
7. Cedzynska. K., Kolacinski, Z. , Izydorczyk, M., and Sroczynski, W., 1991, Vitrification of Hospital Waste Incinerators Residues, *Proc. XIV Int. Syp. Plasma Chemistry*: 2435-2439. Prague.

8. Kolacinski, Z., and Cedzynska, K., 1999, Solidification of Ashes with Thermal Plasma, Proc. Int. Ash Utilization Symposium, Lexington, 6 pages on Ashcd.CEDZYN2. Kentucky.
9. Cedzynska, K., and. Kolacinski, Z., 1997, Destruction of PCB Soil Polluted by Mineral Oil, Proc. XIII Int. Symp. *Plasma Chemistry*: 333-339. Beijing, China.
10. Kolacinski, Z., 1998, Perspectives of Thermal Plasma Application to Waste Treatment - Invited paper for the VIII Int. Symp. *SAP&ETEP 97, Łodz, September 1997*, Post-Conf. Materials: 117-126.

PLASMA TECHNOLOGY IN THE DEACTIVATION OF SPENT ION-EXCHANGERS

Dariusz Dziadko and Iwo Pollo
Department of Chemical Technology, Lublin University of Technology, 40, Nadbystrzycka Str., 20-618 Lublin, Poland

Key words: arc plasma, waste polymers, carbonisation

Abstract: Kinds of plasma applied in the removal of various waste substances are described and diagrams of plasma devices based on the arc discharge phenomenon shown. The advantages and disadvantages of these devices as well as their applications are discussed, problems concerning spent ion exchangers presented and a solution proposed on the basis of the application of an arc plasmatron generating a jet of argon plasma. The thermal stability of the plasmatron and energy distribution over its elements have been tested. Utilisation of a spent ion exchanger at energy ranging from 7.3 to 14.5 kJ per gram of substrate has been attempted and the stability of its operation estimated on the basis of variations in the electrical parameters of the device. The products obtained have in turn been evaluated for mass decrement and degree of charring. Conductivity variations in solutions obtained by mixing treatment products with tap water have also been measured.

1. INTRODUCTION

The application of plasma processes to remove solid substances including wastes is most often practiced on the basis of electrical discharges. The way these are generated determines the properties of the plasma obtained and there by conditions its applications.

While thermal plasma is usually generated in an electric arc the common principle by which the jet is generated does not prevent individual design solutions (Fig.1) from exhibiting significant differences.

Plasma treatment of solid wastes

→ Non - isothermal plasma HPLT High Pressure Low Temperature (corona discharge)
- dedusting on electrostatic filters

→ Mixed plasma e.g. glidarc
- removal of components from gas mixtures

→ Thermal plasma:
- non-oxidizing pyrolysis of pollutants
- melting and vitrification
- recycling of metals
- change in chemical properties and reactivity of materials

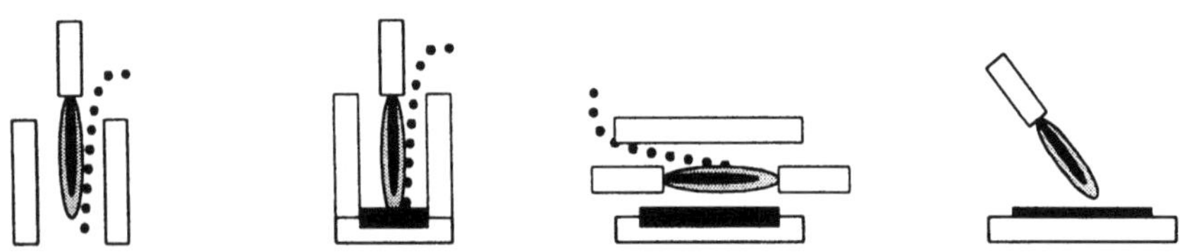

a) b) c) d)

Figure 1. Design diagrams of plasma devices intended for various applications: a) non-oxidising pyrolysis of pollutants; b) melting and vitrification; c) recycling of metals; d) change in chemical properties and reactivity of materials.

As compared to the design solutions of devices operating on the basis of conventional fuels plasma devices are characterised by a number of advantages such as :

– high degree of conversion (high efficiency);
– high temperature of the process ensuring decomposition of volatile substances into simple compounds (very limited re-synthesis of noxious compounds);
– small dimension (low material consumption);
– ease of operation and automation (immediate switch on and off).

However, a few disadvantages of their application also need to be noted, namely:

– high energy consumption (possible retrieval of some energy but in a less valuable form);
– necessary high qualifications of operating staff in case of serious breakdowns which occur easily;

- near-confinement of possible applications to mass production processing.

Despite the mentioned disadvantages, the possibility of applying plasma methods should always be taken into consideration where the processing and utilisation of wastes are concerned. The following are among pioneering plasma applications[1] :
- in metallurgical and steel-making process utilising scrap[2];
- in process that require special conditions such as the vitrification of heavy-metal wastes[3] and nuclear wastes[4] or glass production[5];
- in the destruction of medical wastes[6,7];
- in the obtaining of highly-refined products like fullerenes[8].

While literature data indicate that plasma technologies can be more economical than technologies based on conventional fuels in some concrete applications, attempts at implementation are all too often limited to the pilot scale only. Wider, industrial scale application will require further research into the technological and economic problems.

Our laboratory has a reactor that makes possible the interaction of fine particles of solid matter in plasma. It allows for the attempted deactivation of selected materials that could be noxious and hazardous for the natural environment.

The spent ion-exchangers widely used in the water-demineralisation process include a number of such materials. From the point of view of chemical composition, these are space polymers composed of carbon atoms bound into chain- or ring structures which when substituted with functional groups contain atoms of oxygen, sulfur, nitrogen or phosphorus and are insoluble in water.

The exchangers degrade with time - their operation undergoes progressive deterioration and finally they become wastes. When the wastes come from typical water treatment stations their utilisation is less problematic as they contain metal ions related to the organic skeleton and typical for the natural environment. However, ion exchangers can also serve as carriers of substances that catalyse chemical reactions occurring in the liquid phase and can in such a case contain substances hazardous for the environment. The remnant of their ion-exchange capability can also pose a potential hazard for the environment. Even if they are stored on an adequately prepared dumping ground they can still bind noxious substances from the environment, cumulate them and release them uncontrollably afterwards. The exchangers, as organic compounds, could possibly be burnt together with conventional solid fuels, while pyrolysis in plasma devices offers an alternative solution to the problem.

The mechanism of plasma action on an ion exchanger remains rather unchanged no matter what environment in - neutral, oxidising, or reducing it occurs. In each of the mentioned cases a substance is exposed to the action of high temperature ($5.5 \cdot 10^3$ K or 0.5 eV). From the viewpoint of process realisation it would seem purposeless to apply such extreme temperature conditions. The temperature can be reduced by expanding the plasma volume through a shift in the arc-column application point over the anode surface. An additional advantage of this method is the protection against overheating or damage afforded to the system elements. On the other hand, it can emerge that the interaction between plasma and solid matter involved too short a residence time for particles of the letter in the plasma zone, with the result that there is only partial conversion of the substrate or conversion of some particles only.

The work presented describes the application of plasma treatment in the thermal decomposition of a spent polymer ion exchanger that can be treated as a potential source of graphite.

2. APPARATUS

The pyrolysis of the ion-exchanger was performed in a plasma-chemical reactor with the internal arc depicted in Figure 2

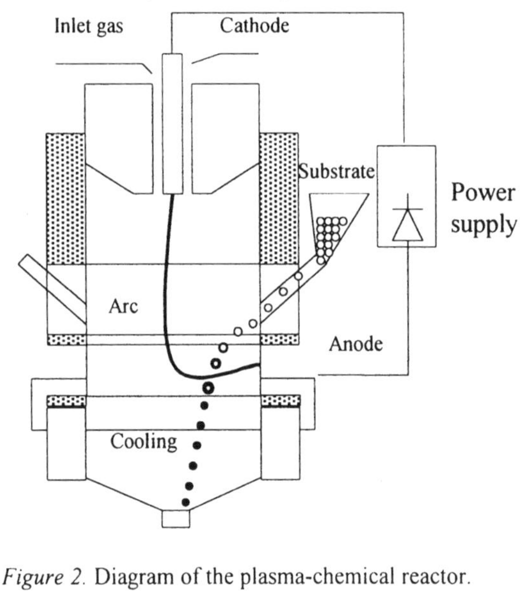

Figure 2. Diagram of the plasma-chemical reactor.

An arc discharge between a cathode with a tungsten insert and a ring-shaped copper anode of ϕ = 48 mm was initiated and maintained in an argon jet from a gas cylinder. The cylinder was connected to adequate devices making smooth control of the gas flow possible.

A DP3a-100 power pack-rectifier of 100 kW power connected to the mains through an induction regulator of the BA 8-2-2 type was applied to feed the plasmatron. Owing to the feeder, autogenous discharge-initiation by means of high-voltage impulses was also possible.

Parts of the reactor were membrane cooled with demineralised water delivered by means of a small hydrophore operating in a closed cycle. The installation was equipped with flow-intensity meters and electronic thermometers.

The anode was protected by a system of coils generating a constant magnetic field to stabilise the arc discharge and induce its rotation around the plasmatron axis.

The electrical parameters of the plasmatron operation were measured and recorded using TZ.21S recorders and an oscilloscope. Substrates were dosed to the reaction zone with a turbine feeder of the Authors' design[9]. A dust chamber and cyclotron were applied to receive the product. Post-reaction gases were disposed of to the atmosphere. To obtain results of greater precision, pyrolysis was performed on the 0.63 mm the grain-size fraction of wastes extracted on sieves. Electrical parameters of the plasmatron were controlled in such a way that the power generated there was of about 8 kW. The substrate was dosed to the reaction zone so that a planned quantity of energy per unit mass of the substrate was obtained.

3. EXPERIMENTAL

The rate of thermal equilibrium becoming established in the plasmatron was tested before the ionite treatment commenced. This was realised by starting the device and measuring the cooling water temperature on individual plasmatron elements. Over the whole power range (7 to 16 kW) the plasmatron was found to arrive at thermal equilibrium in 30 seconds at most.

Checks were also made for heat abstraction out of the cooling system across the power range mentioned. These involved seeing, whether energy abstracted by the coolant from individual plasmatron elements and energy supplied to the system by electrical current were balanced. On the basis of the results obtained (Fig. 4) a satisfactory consistency to the balance can be noted, since the balance difference does not exceed 5%.

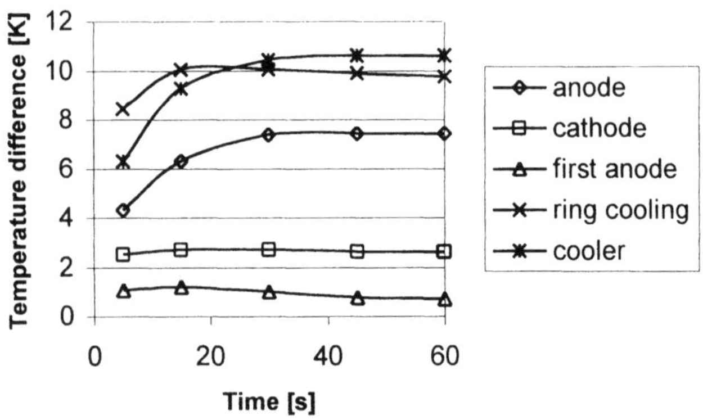

Figure 3. The rate of thermal equilibrium becoming established in the plasmatron.

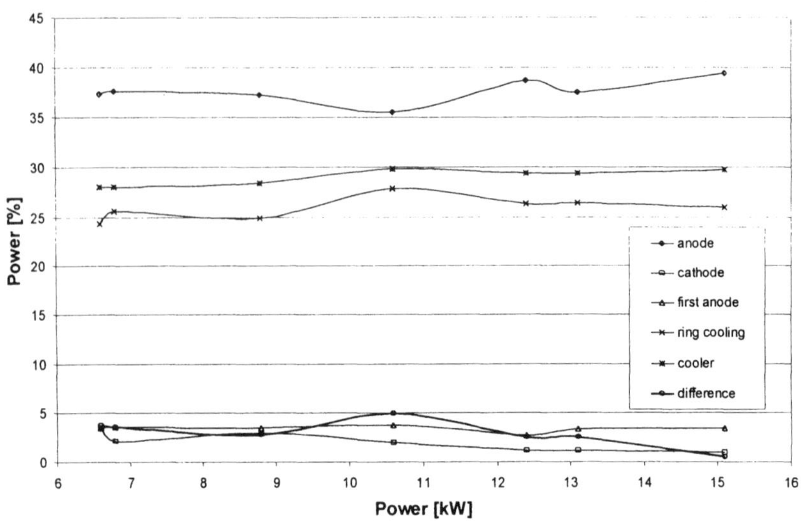

Figure 4. Distribution of energy balanced in plasmatron elements vs. arc discharge power.

A subsequent experimental phase entailed thermal treatment of an ionite such that energy per gram of substrate equalled 9 kJ. The course of the electrical parameters of the process is presented in Figure 5.

A feeder was started at time t = 30 s and substrate was dosed until t = 100 s. The courses obtained for current intensity and voltage curves indicate that the substrate and products of its decomposition deionise the inter-electrode space. This has no effect on the power released in the reactor. The treatment products were charred uniformly and the mass decrement was 48%.

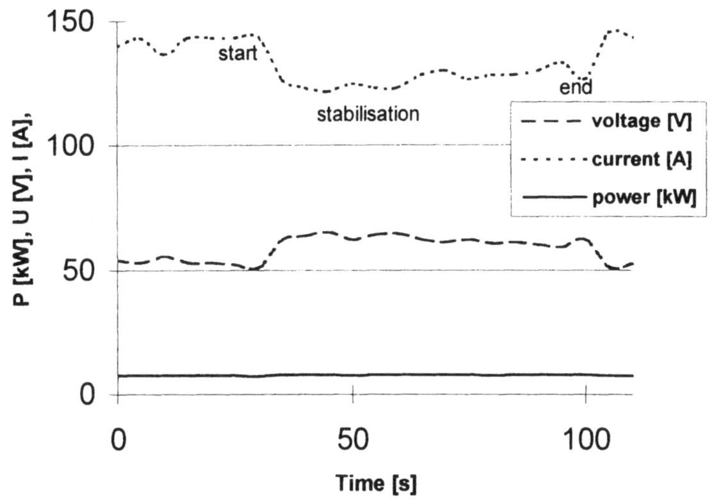

Figure 5. Variations in the electrical parameters of a plasma reactor in the course of treatment of an organic substance at an energy of 9 kJ·g^{-1}.

Ionite decomposition was tested in a similar way at energy of 14.5 kJ g^{-1} per gram substrate. The course of the electrical parameters of the process is presented in Figure 6.

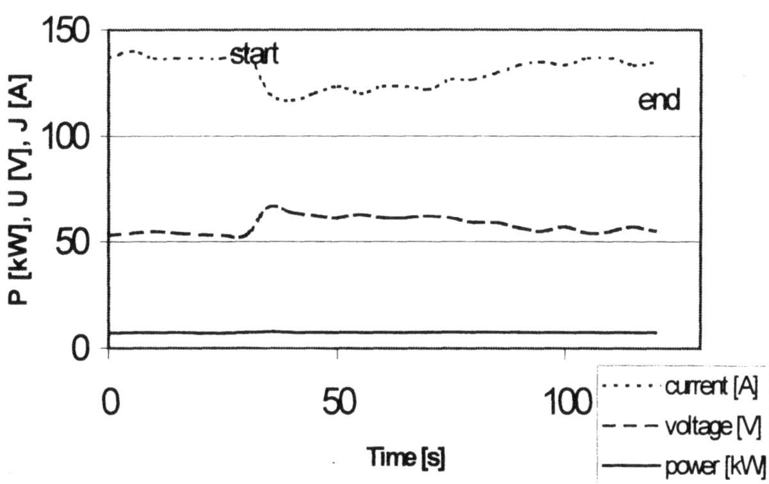

Figure 6. Variations to electrical parameters of a plasma reactor in the course of treatment of an organic substance at energy E = 14.5 kJ· g^{-1}.

In this case, deionisation of the inter-electrode space was only more noticeable in the initial phase of the treatment. The result was complete charring of the resin coupled with a mass decrement of 48%.

A subsequent task was to attempt to determine the minimum energy indispensable of complete charring of a polymer was to be obtained with the treatment at energy expenditures of 7.3 and 8.0 kJ g^{-1}. However in both cases the charring was not complete. Some of the granules were not changed at the surface and the obtained mass decrement ranged from 21 to 40%. It has thus been concluded that the minimum energy expenditure should equal 9 kJ g^{-1}.

A reapplication of the treatment to the partially charred polymer was attempted. The obtained product resembled strongly deformed granules and the mass decrement as compared to the tested exchanger was of 73%.

The carbon products obtained from the experiments were tested for their sorption properties by mixing with tap water and measuring the conductivity of the solutions obtained (Fig. 7).

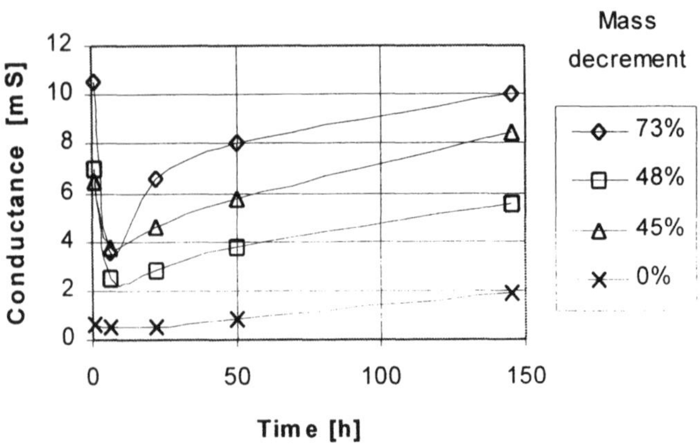

Figure 7. Variations in the conductivity of aqueous solutions in equilibrium with spent ion-exchanging resin and products of its plasma charring differing with respect to their mass decrements.

The conductivity of solutions was high immediately after mixing - a reflection of the washing-out of mineral components. Conductivity then decreased rapidly due to the absorbing action of the carbon. For the sake of comparison a similar test was performed for an unused ionite with the result that continuous diffusion of mineral parts to the solution could be noticed

4. CONCLUSION

1. The plasma device presented can be applied in the utilisation of spent ion-exchangers and arrives at equilibrium in no more than 30 sec.
2. Complete charring of a polymer in the conditions discussed requires the supply of at least 9 kJ of energy per of substrate to the reactor.
3. The carbon obtained by applying the treatment is capable of absorbing mineral substances

REFERENCES

1. Tylko, J. K., 1996 Patent Specification GB 1 596 263, Application No.45839/76 4 Nov.
2. McManus, George J., 1995, Plasma Melting Gets Another Look in the Steel Industry. *Iron & Steel Engineer, Vol. 72*: 49-50
3. Cortez, R., Zaghloul, H. H., and Stephenson, L. D., 1996, Laboratory Scale Thermal Plasma arc Vitrification Studies of Heavy Metal-laden Waste. *Journal of the Air & Waste Management Association, Vol. 46*: 1075-80.
4. Filius, K. D., and Whitworth, C. G., 1996, Emissions Characterization and Off-gas System Development for Processing Simulated Mixed Gaste in a Plasma Centrifugal Furnace. *Hazardous Waste & Hazardous Materials, Vol. 13*: 143-52.
5. Bentley, J. A., 1996, Pilot Scale Furnace for the Demonstration of Transferred Arc Melting. *Glass Technology, Vol. 37*: 192-194
6. Rozporządzenie Ministra Transportu i Gospodarki Morskiej Dz. U. z dn. 25 czerwca 1999r. No 57, poz. 608.
7. Lapierre, R., 1996, Zapping the red out of red bag waste. *Electrical World, Vol. 210*: SS3-SS4.
8. Huczko, A., Lange, H., and Płotczyk, W., 1997, Decomposition of Spent Ion Exchanger in Arc Discharge Plasma. *Plasma Chemistry IV*. Lublin.
9. Dziadko, D., and Pollo. I., 1997, Searching for Optimum Design of a Feeder for Solids in a Laboratory Arc Plasma Reactor. *Chemical Processes In Low Temperature Plasma*: 85-89. Wydawnictwa Uczelniane Politechniki Lubelskiej, Lublin (in Polish).

PLASMA GASIFICATION OF SURROGATE AND REAL WASTE PLASTICS

Andrzej Huczko and Maciej Sioda
Laboratory of Plasma Chemistry, Department of Chemistry, Warsaw University, 1, Pasteur, 02-093 Warsaw, Poland

Key words: gasification, plasma pyrolysis, PE, PET, waste plastics

Abstract: Model (standard polyethylene PE) and real plastic waste (polyethylene terephthalate PET and polyethylene PE bottle scraps and polyethylanated carton scraps) were decomposed in a carbon arc plasma. Total conversion of reactants into highly unsaturated products (mostly acetylene) was achieved with some additional hydrocarbons being formed from the evolving hydrogen and carbon vapours resulting from graphite electrode sublimation.

1. INTRODUCTION

The majority of domestic wastes are currently disposed of to landfills world-wide (excluding the well-developed countries in which incineration has become the best solution). However, this practice becomes unpopular due to the shortage of suitable landfill sites and the introduction of more stringent environmental laws. Due to the more and more strict legal regulations even the treatment of wastes which normally have been disposed to the landfills becomes a necessity now. Thus many attempts have been made on the development of new techniques and processes, which can reduce the volume of the final product and decrease its environmental impact. Thermal plasma is one of the possible solutions, since plasma systems are simple to operate and can be used for large scale operations.

Moreover, the harmful wastes (e.g. halogenated chemicals) can be effectively processed since the reaction temperatures (within at least a few thousands K) allow for the total atomisation of reactant structure. Thus, the thermal plasma technology is considered to be an excellent method to solve the problems of the recycling of special industrial wastes and hazardous materials. This technique is matured with many examples of various pilot and full-scale installations world-wide (e.g. refer to Figs 1–2).

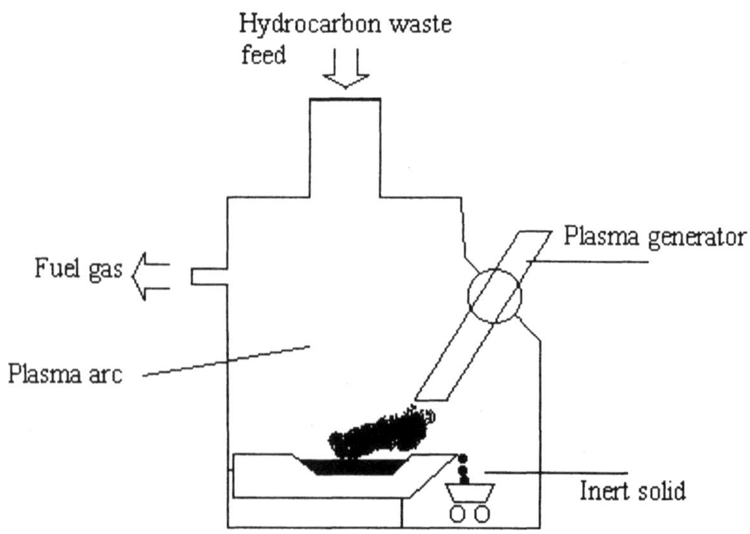

Figure 1. Thermal plasma furnace for processing of hydrocarbon waste[1].

The treatment of the oxidic dusts to recycle the valuables and immobilise the harmful substances demands high temperatures and non-oxidising atmospheres. These demands are met by plasma melting and reduction using a D.C. carbon arc plasma furnace. The ferroalloys are produced from the metallurgical dusts (1–4 tons/h) at the power level 2–7 MW and reaction temperatures within 1350–1500°C. Pt group metals are effectively recovered from spent car catalysis using plasma systems (power level 2MW - Fig 2).

Municipal solid waste (MSW) and other (clinical, radioactive) waste are processed in pilot plasma plant (power 150kW. Various wastes (e.g. MSW, chemical, coal, hospital) can be converted into a syn-gas (CO/H_2 mixture) by addition of water vapour into the plasma zone.

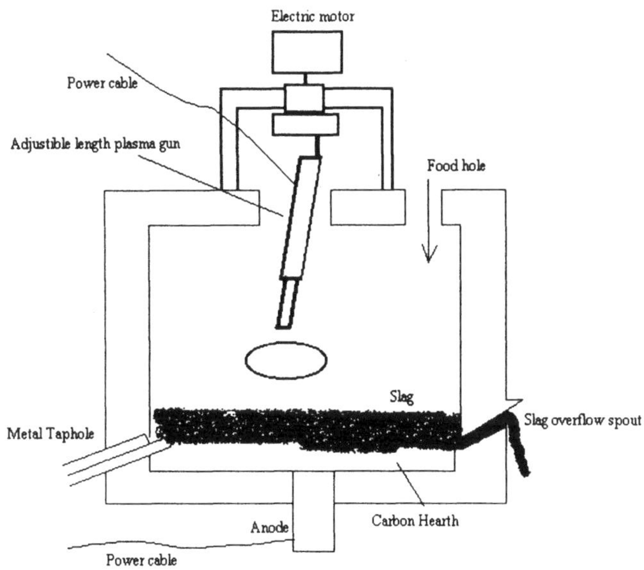

Figure 2. PGM plasma furnace for the treatment of spent car catalysts[2].

Table 1 shows (as of 1992) the active plasma projects related to the environment protection. By now the list of the plasma projects under way has been increased. PLASMOX, PLASMAREC, and PLASMARC processes for the treatment of chemical toxic wastes, recovery of metals and the melting of nuclear wastes, respectively, have been commercialised in Germany and Switzerland[3]. The destruction of ozone–depleting substance (ODSs) using the PLASCON thermal plasma technology is well under way in Australia[4]. The plasma melting of the MSW incinerator ashes is also well advanced in Japan[5].

Table 1. Active plasma projects.

Designer	Waste	Capacity	Plasma	Electrode	Site
P.E.C.	MSW		Nitrogen	Dc trans, arc torch	Raleigh, S. Carolina
P.E.C.	MSW	0.2 tons/h	Nitrogen	Dc torch	Canada, Resorption
Arc Technology	PCB	1.5 tons/h		Dc graphite	Pilot plant USA
PYRAL	PCB			Ac-graphite	Pilot plant USA
Krupp Mak AG Germany	Flyash and filter dust	1.28 tons/h	Argon-nitrogen	3-phase ac torch	Demo plant Hamburg
Daido Japan	MSW	0.8 tons/h		Ac-graphite	
Retech with	Mixed	200 lbs/h	Air	Dc torch	Pilot plant

Designer	Waste	Capacity	Plasma	Electrode	Site
DOE	radioactive and organic wastes				Butte, Montana
Retech Inc. Calif. & MGC Plasma	Toxic solids and liquids	1.1 tons/h	Air	Dc torch	Pilot plant Muttenz Switzerland
Retech with EPA	Organic liquids	120-500 lbs/h	Air	Dc torch	Pilot plant Butte, Montana
Aerospatiale	Organic liquids	439 lbs/h	Air	Dc torch	Rhone-Poulenc Grenoble
BS/TRD	Alloy steel EAF dust	1.5 tons/h	Argon	Dc graphite	British Steel Sheffield, UK
ILSERV/TRD	Alloy steel EAF dust	5 tons/h	Nitrogen	Dc graphite	ILVA Italy
IMS/TRD	Carbon steel EAF dust	1.5 tons/h	Nitrogen	Dc graphite	Florida Steel Tennessee
IMS/TRD	Carbon steel EAF dust	3 tons/h	Nitrogen	Dc graphite	Nucor-Yamato Steel Blytheville. Ark
Davy McKee	Carbon steel EAF dust	1 tons/h	Argon	Dc torch	Stockton, UK
Mintek	Carbon steel EAF dust	0.5 tons/h	Argon	Dc graphite	S. Africa
Scandust	Carbon steel EAF dust	5 tons/h	CO	Dc torch	Landskrona, Sweden
Westinghouse	Organic liquids	180 gal/h	Oxygen	Dc arc	Mobile plant Rio de Janeiro, Brazil
Westinghouse	Toxic landfill	2.5 tons./h	Air	Dc arc	Pilot plant Madison, PA
Skygas	Toxic Waste	1.0 tons./h		Ac graphite	Montana, USA
Aerospatiale	Hospital Waste	0.5 tons./h	Air	Dc torch	CNIM
Westinghouse	Scrap Metal	50 tons/h	Air	Dc arc	GM Defiance, Ohio

Concerning the processing of MSW in conventional incinerators the rising problem is related to the possible formation of polychlorinated dibenzo-p-dioxins (PCDDs) and dibenzofurans (PCDFs) as reviewed by Tupprainen et al.[6].

Moreover, the burning of waste plastics is the final energy recirculation since all 'chemistry' of the reactants is irreversibly lost. Plasma processing can overcome those drawbacks, mostly due to the high reaction temperatures and neutral atmosphere applied.

In the field of conventional thermal processing of waste polymers, a large number of studies have been carried out on the decomposition of plastics in various experimental systems at elevated temperatures[7,8,9]. The conversions obtained are high but the economics of the proposed reaction routes is doubtful, mostly due to complex composition of reaction liquid and gaseous products.

Considering the annual production of MSW in Europe (2,8 billion tons) and the data from Figs 3 and 4 it is evident that the new techniques to process spent plastics are urgently sought (Fig 5).

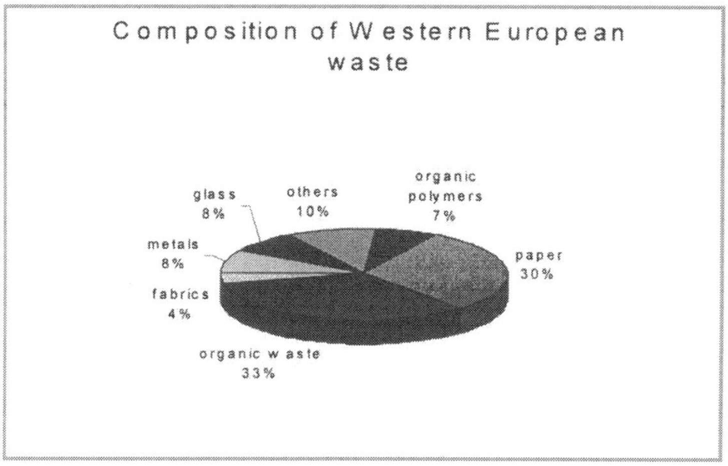

Figure 3. Composition of Western European Waste.

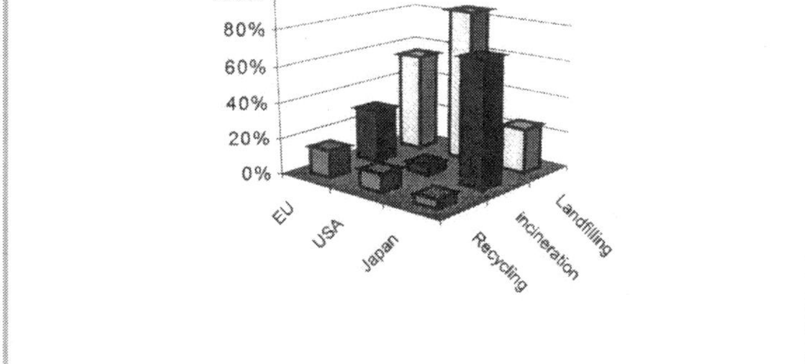

Figure 4. Percent fractions of currently applied utilisation methods

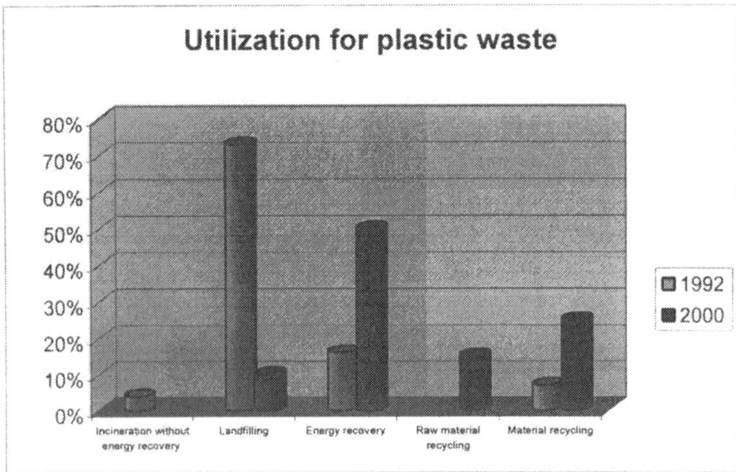

Figure 5. Current and prognoses techniques for the utilisation of plastic waste.

The plasma systems can offer an interesting solution for the recycling of spent polymers.

In this study we present the results of the exploration study on a plasma pyrolysis of polymers. Since acetylene is routinely the main product of the plasma decomposition of carbon–bearing reactants[10], the reducing atmosphere of carbon arc has been chosen for the process since it may favour the additional production of unsaturated hydrocarbons formed from the hydrogen (resulting from the decomposition of polyolefines) according to the scheme:

$$(CH_2)_n + nC \rightarrow nC_2H_2$$

2. EXPERIMENTAL

The tests were carried out in the experimental system presented in schematically Figure 6.

The details of the apparatus and the spectroscopic arrangement have been presented elsewhere[11,12]. This system has been routinely used in the synthesis of carbon nanostructures (fullerenes and carbon nanotubes)[13]. Polyethylene (PE) was selected as a model reactant since it may constitute above 50% of all waste plastics (Fig 7).

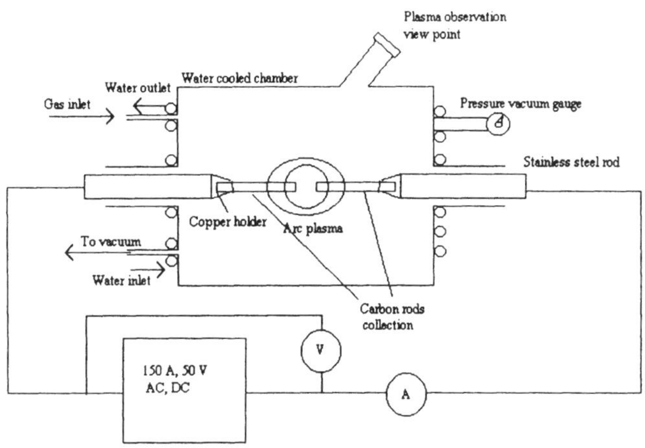

Figure 6. Experimental system.

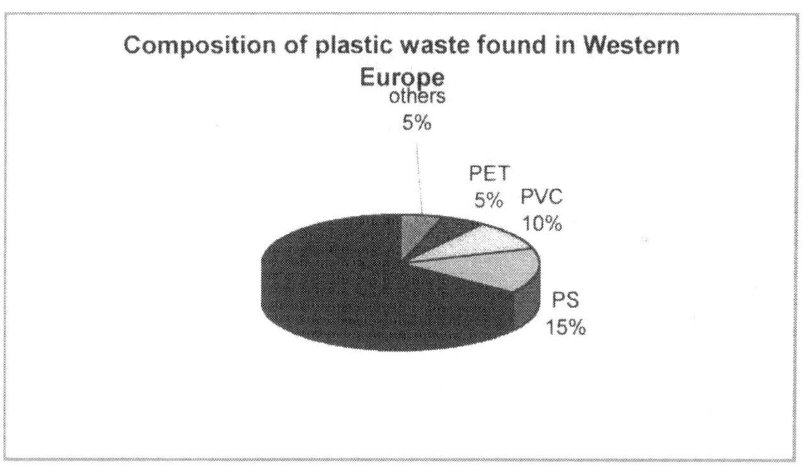

Figure 7. Composition of plastic waste found in Western Europe.

The pyrolysis was carried out under static conditions in neutral He, Ar and reactive H_2 gases at medium pressure. The consumable graphite was packed with solid reactant and carbon vapours resulting from electrode sublimation provided for the reductive reaction environment. Both solid and gaseous products were collected and analysed (GCMS), and the mass balance was performed to evaluate the conversion of reactants.

3. RESULTS AND DISCUSSION

Three sets of experiments were carried out: (i)graphite anode sublimation under hydrogen atmosphere; (ii) model PE pyrolysis under various operational parameters; (iii) decomposition of "real" plastic waste (spent PE and PET, PE – covered carton).

The graphite anodes were burned out in the hydrogen atmosphere to verify the reaction channels related to direct hydrogenation of sublimated carbon. The process variables are shown in Table 2 while Table 3 contains the experimental results (composition of off–gas and C/H conversion to gaseous products).

The presented data prove that the carbon (originating from the anode) can be converted under hydrogen presence into unsaturated products, mostly acetylene. The results of spectroscopic diagnostics of the plasma zone showed only the presence of carbon species and hydrogen atoms which is reasonable considering the temperatures in the interelectrode gap (see Fig 8). During the expansion of hot plasma the quenching processes occur with the formation of CH radicals, which finally form C_2H_2. Higher content of ethylene in test No. 3 can be related to higher pressure which favour the synthesis of more complex molecules.

Table 2. Operational parameters for graphite arc sublimation in hydrogen atmosphere.

Exp. No.	H_2 content in H_2/He mixture (vol%)	Initial pressure (mb)	Operational pressure (mb)	Final Pressure (mb)	Arc Current (A)	Arc Voltage (V)	Anode erosion rate (mg/s)
1	50	102	155	106	95	17,5	3.2
2	100	88	140	92	102	22.5	4.6
3	100	582	1000	889	85	40	12.3

Table 3. Product composition

Exp, No	Gaseous products vol %			C and H (from PE) conversion to gaseous products %	
	CH_4	C_2H_4	C_2H_2	C	H
1	0.16	0.06	0.56	14.4	18.5
2	0.05	0.14	0.91	34.4	14.9
3	0.06	1.85	1.19	26	24.5

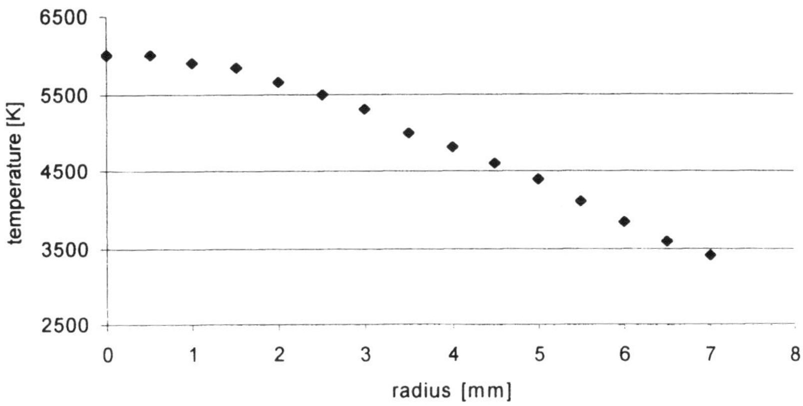

Figure 8. Arc plasma temperatures within the interelectrode gap.

Thus, the presented results showed that carbon sublimating from the anode can react with hydrogen (if present in the reaction zone) to form mostly acetylene.

Tables 4 and 5 present the operational parameters and analytical results to model (standard) polyethylene PE conversion in the arc plasma.

Table 4. Operational parameters for PE pyrolysis in the arc discharge.

Exp. No.	Mass of processed PE (g)	Plasma Gas	Initial pressure (mb)	Operational pressure (mb)	Final Pressure (mb)	Arc Current (A)	Arc Voltage (V)	Anode erosion rate (mg/s)
4	0.1190	He	80	150	89	97	17	6.2
5	0.2448	He	73	152	95	95	17	5.0
6	0.4924	He	92	192	153	85	26	5.9
7	0.4818	He	3	89	51	100	20	13.7
8	0.5364	He	30	145	82	102	22	10.2
9	0.5013	He	93	190	150	827	21	6.2
10	0.4299	He	603	960	623	777	22	3.2
11	0.4847	Ar	82	310	159	86	32	9.5
12	0.5859	Ar	81	260	149	80	26	4.7

Table 5. Product composition and reactant conversion for PE processing in the arc plasma.

Exp, No	Gaseous products vol %			C and H (from PE) conversion to gaseous products %	
	CH_4	C_2H_4	C_2H_2	C	H
4	0.08	0	0.85	213.9	121.3
5	0.36	0.43	0.60	111.1	99.9
6	0.07	0.05	2.37	104.7	55.4

	Gaseous products vol %			C and H (from PE) conversion to gaseous products %	
7	0.08	0.07	0.59	35.4	22.3
8	0.04	0.26	1.93	83.1	47.8
9	0.11	0.10	2.05	95.3	52.3
10	0.32	0.39	1.54	95.7	65.2
11	1.00	0.03	1.30	68.9	75.0
12	0.10	0.09	1.10	30.2	42.0

Regardless the experimental conditions PE was totally decomposed with gaseous products being formed during the quenching stage. No CH radicals were detected in the plasma zone, thus the products (mostly acetylene) result from recombination processes. Some aromatic compounds (mostly benzene) were also detected in off-gas in minor quantities. The variations of operational parameters did not seem to influence distinctly the product composition, probably due to the excessive energy available from the plasma in relation to the mass of processed reactant. C and H conversions higher than 100% confirmed the reaction channel sought, e.g. the additional formation of unsaturated hydrocarbons from carbon vapours (from the anode) and the residual hydrogen (see scheme 1).

Several experiments have been carried out with the real plastic wastes (PE and PET bottle and polyethylenated carbon scraps) with some exemplary results shown in Table 6 (operational parameters) and Table 7 (analytical results). Both qualitative and quantitative results of testing are much the same comparing with the processing of model polymer (PE).

Table 6. Operational parameters for real plastic waste plasma processing (*plasma gas – He*).

Exp. No.	Processed reactants	Mass of processed PE (g)	Initial pressure (mb)	Operational pressure (mb)	Final Pressure (mb)	Arc Current (A)	Arc Volt. (V)	Anode erosion rate (mg/s)
13	PET bottle scraps	1.4696	88	220	135	75	25	9.9
14	PE bottle scraps	0.9786	76	285	180	70	36	5.5
15	Polyethylene covered carton scraps	0.7704	95	242	169	85	20	5.2
16	As above	0.9330	654	900	675	70	27	2.8

In conclusion one can say, that waste plastics can be effectively processed in the carbon arc plasma with the total decomposition to the gaseous unsaturated products (mostly acetylene). Following the transformation of PE into acetylene the evolving hydrogen vigorously reacts with carbon vapours (from the sublimated anode) and forms additional unsaturated hydrocarbons in the system.

Table 7. Product composition and reactant conversions (real plastic plasma processing).

Exp. No	Gaseous products vol %			C and H conversion to gaseous products %	
	CH_4	C_2H_4	C_2H_2	C	H
13	0.32	0.62	0.92	36.5	29.3
14	0.07	0.19	1.75	37.2	21.5
15	0.55	0.25	1.01	49.2	45.4
16	0.95	0.07	3.00	112.9	141.0

REFERENCES

1. Ibid., S. L. Camacho, P. 7.4.
2. Williams, J. K. , 1994, High Temp. Chem. Process.: 3, 707.
3. Hoffelner, W., private communication.
4. Murphy, A.B., 1997, *J. & Proc. Royal Society* NSW, 130, parts 3-4: 93-108.
5. Matsumaru, K. , Fukuyama, H., Susa, M., Nagata, K., 1998, Thin Solid Films, 316: 105-110.
6. Tuppurainen, K., Halonen, I., Ruokojarri, P., Tarhanen, J., and Ruuskanen, J., 1998, *Chemosphere*, 36, nr.7: 1493 – 1511.
7. Conesa, J. A., Font, R., Marcilla, A., Garcia, A.N. , 1994, *Energy & Fuels*, 8:1238 - 1246.
8. Cozzani, V., Nicolalla, C. , Rovatti, M. , and Tognotti, L., 1997, *Ind. Eng. Chem. Res.*, 36: 342 – 348.
9. Ramdoss, P.K. , and Tarre, A. R., 1998, *Fuel*, 77, No 4: 293 - 299.
10. Laflamme, C.B., Jurewicz, J.W., Gravelle, D. V., and Boulos, M. I., 1990, *Chem. Eng. Sci.* 45. No,8: 2483 - 2487.
11. Lange, H., Baranowski, P. , Byszewski, P., and Huczko, A., 1997, *Rev. Sci. Instr.*, 68, No. 10: 3723 – 3727.
12. Lange, H, Saidane, K., Razafinimana, M., and Gleizes, A., 1999, *J. Phys. D: Appl. Phys.*, 32: 1024 – 1030.
13. Huczko, A., Lange, H., Byszewski, P., Popławska, M., and Starski, A., 1997, *J. Phys. Chem. A*, 101, No. 7: 1267 – 9.

ABOUT THE EDITORS

LUCJAN PAWŁOWSKI

Lucjan Pawłowski, was born in Poland, 1946. He studied chemistry at the Maria Curie-Skłodowska University. He got his Ph.D. in 1976, and D.Sc. (habilitation) in 1980 both at the Wrocław University of Technology. In 1986 the President of Poland nominated L. Pawłowski to a full professor. He started research on application of ion exchange for water and wastewater treatment. As a result he together with B. Bolto from CSIRO Australia, has published a book „Wastewater Treatment by Ion Exchange" in which they summarised their own results and experience of ion exchange area. In 1980 L. Pawłowski was elected President of International Committee „Chemistry for Protection of the Environment". He was Chairman of the Environmental Chemistry Division of the Polish Chemical Society in 1980-1984. In 1994 he was elected the Deputy President of the Polish Chemical Society and in the same year, the Deputy President of the Presidium Polish Academy of Science Committee „Men and Biosphere". In 1999 he was elected as a President of the Committee "Environmental Engineering" of Polish Academy of Science. He is member of editorial board of international journals: „Reactive Polymers", „The Science of the Total Environment", „Environment International", and „Journal of Ecological Chemistry". In 1991 he was elected as the Deputy Rector of the Lublin University of Technology, and this post he has hold for two terms (1991-1996). Currently he is Director of Environmental Engineering Institute. He has published 19 books, over 128 papers, and authored 68 patents.

MARZENNA R. DUDZIŃSKA

Marzenna R. Dudzińska received M.Sc. in chemistry in 1983 from Marie Curie-Skłodowska University in Lublin, Poland. In 1983-1988 she was employed as research worker and teacher at Department of Physical Chemistry at M.Curie-Skłodowska University. In 1987 she worked in Institute of Spectroscopy and Applied Spectrochemistry in Dortmund, FRG on the project combining chromatography and spectroscopy. In 1988 she was hired by Lublin University of Technology, Poland, and worked for Water and Wastewater Treatment Department. She got a Fulbright Scholarship in 1989, and spent 1989/1990 at University of Houston, Texas, in Civil and Environmental Engineering Department. She received Ph.D. in environmental chemistry from Marie Curie-Skołodowska University in 1992. Now she works for Institute of Environmental Protection Engineering at Lublin University of Technology. She has authored numerous papers related to adsorption and ion exchange processes and complexation mechanism in water systems. Since 1996 she has been working on projects sponsored by Polish Committee for Scientific Research on waste incineration, PCDF/D emission and PCDF/D formation/degradation. She was awarded by Polish Chemical Society for her master thesis in 1983, by Polish Ministry of Science and Education in 1987 and1992, and by Rector of Lublin University of Technology in 1994, 1997 and 1999. She is a member of Polish Chemical Society, and International Committee "Chemistry for Protection of the Environment".

MARJORIE A. GONZALEZ

Marjorie A. Gonzalez is the Environmental Assurance Manager for the Defense and Nuclear Technologies Directorate at Lawrence Livermore National laboratory (LLNL). She currently is responsible for providing environmental technical and management advice to researchers, with her primary focus being pollution prevention efforts. She has also been a principal co-investigator on an environmental projects for the Middle Urals' region of Russia, collaborating with the environmental scientists from the Institute of Industrial Ecology of the Russian Academy of Sciences located in Yekaterinberg, Russia and scientists from the Russian Federal Nuclear Center Institute of Technical Physics located in Snezhinks, Russia.

An LLNL employee for 20 years, Ms. Gonzalez's work has covered a broad spectrum of environmental issues, but has always focused on cost-effective integration of programmatic goals with technical, policy, and regulatory requirements. She has authored numerous articles related to on-line detection of pollutants in wastewater and the development and implementation of source reduction and recycling technologies. Ms. Gonzalez received her Bachelor of Science degree in Mathematics from California State Polytechnic University, San Luis Obsipo, California, USA and her master of Science degree in Environmental Engineering from the University of Cincinnati, Ohio, USA. She is Qualified Environmental Assessor certified by Institute of Professional Environmental Practice.

AUTHOR INDEX

SUBJECT INDEX

FSC
www.fsc.org
MIX
Papier | Fördert
gute Waldnutzung
FSC® C083411

Kolibri 360 GmbH
Ferdinand-Jühlke-Straße 7
99095 Erfurt, Deutschland
produktsicherheit@kolibri360.de